WILDFLOWERS
OF THE
COAST

Word Portraits of Selected Flowering
Plants & Shrubs of the Pacific Northwest

By

JOHN G. FITCH

With illustrations by Bonnie Moro

2013

For information, please contact:
Lucretius Press, c/o John Fitch
Department of Greek & Roman Studies,
University of Victoria, PO Box 3045,
Victoria BC, Canada V8W 3P4.

First printing 2013

ISBN 978-0-9920358-0-8

Contents

Part 1: SEA COAST

Part 2: FIELD AND SLOPE

Part 3: FOREST

Part 4: MOUNTAINS

Illustrations

Part 1

Part 2

Part 3

Part 4

Introduction

The wildflowers in this volume are arranged for convenience in four environmental sections: seacoast, meadow and slope, forest, and mountain. In the real world, of course, many plants cross these boundaries and grow happily in more than one habitat. Within each of this book's four sections, the plants are described approximately in the order of flowering. Again this ordering is rough-and-ready in comparison with nature's infinite variety and adaptability, particularly in the mountains where many plants bloom simultaneously in the short flowering season.

This volume describes some 70 plants—a relatively small selection. It does, however, include many of the most familiar and best-loved plants of the Pacific Northwest. One criterion for inclusion was whether the plant seemed to me to offer the potential for a poem. This depended on my own experience and familiarity with the plant, on my response to its form and colour. Knowledge, after all, is always personal. But 'personal' is not meant here in an exclusive sense. If these poems work at all, I hope that readers familiar with particular plants will find something truthful in my descriptions, while those unfamiliar with the plants will be encouraged to get to know them.

But why write these descriptions as poems? Because poetry is our best and fullest kind of speech or writing —the most memorable and most vivid, encompassing not only thought but also emotion. Wildflowers are such miracles of adaptation, of form and colour, that they deserve the best words we are capable of, just as they have long inspired painters to use all their artistry in portraying them. As flowers represent one aspect of nature's creativity, poetry represents another; it seems appropriate that the one should celebrate the other.

The possibility of conveying botanical information, in addition to aesthetic appreciation, through poetry demands a further comment. In the last century poetry unhappily become divorced from systematic and scientific knowledge of all kinds. This divorce was due in part to the notion that systematic knowledge must be objective and impersonal. In fact such knowledge, however precise, is not objective: it is created by human beings, from a certain point of view, often by imaginative leaps of thought, with passion and excitement. These are characteristics of human thinking that are not at all alien to poetry.

There is, in fact, an immensely long tradition of poetry married to bodies of knowledge, running from Homer and Hesiod in the eight century B.C., to Lucretius' magnificent *On the Nature of Things* in the first century B.C., to James Thomson's popular *The Seasons* in the eighteenth century, and beyond. Admittedly the relationship between poetry and knowledge is constantly in need of renewal, as knowledge changes and advances with dizzying speed. But perhaps the current divorce is only temporary.

Acknowledgements

In any field of knowledge, we stand on the shoulders of giants. In writing about wildflowers of this region, I think particularly of Lewis Clark and Nancy Turner. Where a phrase from these or other writers seemed to me perfect, and appropriate to my poem, I have not hesitated to adopt or adapt it—a procedure familiar in the tradition of knowledge-based poetry just mentioned. Such borrowings are a tribute, an acknowledgement that something has been said with such precision or beauty that it could not be improved—but could perhaps take on new life in a new context. I have aimed to identify such borrowings in the endnotes to the poems; if any have been omitted, I hope the oversight will be forgiven.

My thanks are due to Dr Nancy Turner, my former colleague at the University of Victoria, who—with a generosity familiar to those who know her—agreed to check my poems for botanical and ethnobotanical accuracy. I am grateful to Barbara Colebrook Peace and to Peter Sanger for encouragement along the way. Many students in my courses at UVic responded imaginatively to the concept of 'poetry of knowledge', and were adventurous enough to compose their own poems in this genre.

My teacher in identifying both flowers and birds was my wife Linda; she learnt from her father Sidney Smith, who in turn had learnt from his father James. So, like runners, we pass on the torch of knowledge.

'Vanilla Leaf', 'Sedum', 'Twinflower', 'Trillium' and 'Silky Phacelia' appeared previously in *The Antigonish Review*; 'Devil's Club' in *Chiron Review*; 'Arbutus' in *The Victoria Naturalist*; 'Paintbrush' in *Manzanita: Poetry & Prose of the Mother Lode & Sierra*.

Overture

Saltwater deeps, fish-teeming, clear as crystal,
edged by sharp rocks or shingle beaches,
hunted by osprey and eagle, haunted by gull-cries;
meadows and slopes, sweetly scented
by sun-warmed grasses and pine and fir;
forests that rise in waves from the sea,
softening the rugged forms of the land;
and always mountains ringing the view,
holding their snows till June or all summer—
this is our home, the place that enfolds us
and shapes our lives, a nurturing mother.
And in every quarter wildflowers riot
—on spray-dashed rocks, in grassy clearings,
in woodland shade and alpine meadows—
clothing the earth, a brilliant blanket
to shield her body from rain and sun.
They have evolved a special fitness
for this one place, adapting fully
to its summer droughts and winter rains.
By learning to know them, to understand
the way their growth is matched to the seasons,
to the soils and lie of the land, we too
grow into connection with soil and water,
becoming natives, rooted here.

Easter Lily, *Erythronium oregonum*

Part I:
SEA COAST

Satinflower, *Sisyrinchium douglasii* (*Olsynium douglasii*)

They grow on age-worn rocks
in cracks and crevices, pockets
of scant and gritty soil.
Sheathed in grassy stems
the lustrous flowers slip out
to challenge blustery March;
smooth and sheer as satin,
red-purple with anthers of gold.
Even in overcast, their sheen
glistens: close-up you see
not colour so much as light,
or colour mingled with light.
In rain they bend
like bright umbrellas to shed the drops.
With dancing lightness
they survive the squalls and showers
that flatten the foreign crocuses.

Rusty-haired Saxifrage, *Saxifraga rufidula*

This saxifrage is tiny,
no higher than the length of your thumb,
but it stands as an emblem of toughness.
It lives in fissures in the rocks,
soaking in spring, in summer
baking, in winter frozen.
Its leaves are stiff and leathery,
glossy green above
but reddish-brown beneath;
tell them by the teeth at their edges,
like the regular cogs of a gear-wheel.
The lilliputian stem
carries small exquisite flowers
whose fresh whiteness contrasts
with the dark-wet mosses beneath.
Unlike other saxifrage flowers
they have no speckles of colour
on the petals: those spots you see
are the rounded reddish anthers.
In summer's drought, a marvel:
the leaves curl up, displaying
the brown of the underside,
looking utterly scorched and withered;
but with rain and cloudy weather
they reverse themselves, and show
their lively green once more—
resilient, born survivors.

Saxifraga means 'rock-breaker':
because the genus grows
in rocky clefts, it was thought
to cleave the rocks with its roots,
as if with the strength of trees.
Rufidula, 'the little
red-haired one', refers
to the hairs on the back of the leaves.
But if species names described
the nature of the plant, *tenacula*
would have suited: 'the tough little one.'

Easter Lily/White Fawn Lily, *Erythronium oregonum*

White drifts at Easter in the grass
or on broken rocky slopes:
a reminder of snows just past,
and a chaste annunciation
of the great flowering season.
From the bulbs grow fleshy leaves,
dappled like the skins of fawns.
The flowerstems bend into shepherds' crooks,
displaying the sturdy backs of the petals
but averting the purer white of the face.
Slowly the petals curve back
and yet resist that bending,
so the flower seems filled with a lively
tension, like that of the dance.
The face, if you lift the flower,
is lit by golden anthers
and by tongues of orange at the centre
like pentecostal flames.
          Beauty brings its dangers:
I have seen girls at Easter
with handfuls of these white lilies.
If the flower is picked, it cannot seed;
if the leaf is picked, the bulb will die.
Let us teach our kids to admire them,
and not to tear their heads off
--to love them, not commandeer them.
But the greater threat are the captains
of our sprawling, despoiling culture,
who scour away wild beauty
for barren parking-lots
and rows of houses with authorised
plants in authorised places.

If you would grow the lilies
to gladden the spring, and replace
a few of the myriads lost:
sow the fresh seeds (so golden!)
in flats, for good germination.
Next spring they will show like blades
of grass: plant them out in groups,
in the grass or under trees.
Each year the leaves grow broader.
Be patient, in accord with Nature
who does not create such marvels in haste:
perhaps in five years, the flowers will come.

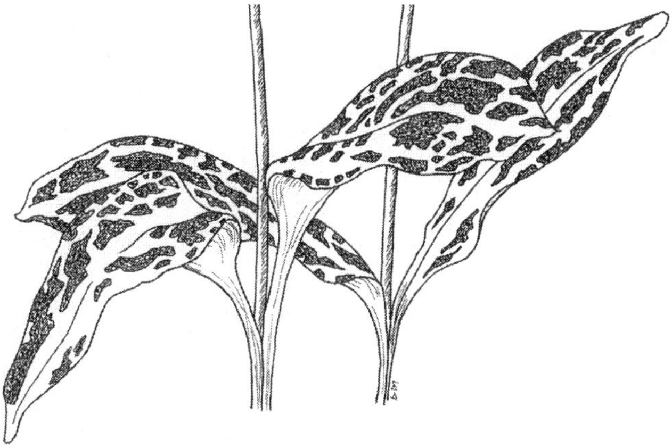

Monkey Flower
(smaller, *Mimulus alsinoides*; larger, *M. guttatus*)

In hollows among the rocks
and where the moss holds moisture
you discover the tiny flowers
cheerful and golden-yellow
of *mimulus*, 'the little mimic,'
aping monkey faces.
With a magnifying glass
you can study the puzzling marks,
maroon and scarlet, on the grinning
lower lip of the face,
which are matched by spots of red
on the calyx that holds the flower.
But where there is plentiful water,
a seepage, or a bright trickle
over glistening, blackened rocks,
there you will find the showier
faces of the larger species,
a golden gamboge yellow,
glowing with the sun's own brilliance.
Again the protruding lower
lobe is "speckled" (*guttatus*),
but with larger drops of red
that need no magnifying.

These are plants of the light:
they live on sun, air, water.

Mist Maidens, *Romanzoffia tracyi*

Along these stormy shores,
on misty Haida Gwaii,
on the outer coast of the Island,
where so many ships have been tossed
and broken by winter's fury,
this exquisite plant is cached
among the spray-wet rocks
like treasure-trove from a wreck.
How does it survive? With a glossy
wax-like covering on the leaves,
to repel the damaging salt.
Its leaves are broader than long,
dark green, with edges cut
into five lobes or more;
the flowers held slightly above them,
clear white against their darkness.
You might think it a saxifrage
at a glance, but the saxifrages
keep their flowers' five petals
distinct, whereas this plant
fuses its petals at the base,
creating a shallow funnel.
The unfused tops of the petals
form lobes around the flowers,
which echo the rounded lobes
of the leaves directly below them.
Such elegance of design!
Count it a Romanzoff treasure
shipwrecked and tossed ashore
and glistening there on the margins,
the very verge of the land.

Sea Blush, *Plectritis congesta*

Sea Blush: carpets of pink,
clear, intense and vivid,
on slopes above the sea.
Here the soil is thin,
and each of the flowering stems
just two or three inches high.
Their compact domes consist
of a cluster of tiny flowers;
there may well be thirty, 'plaited'
(*plêctros*) and 'congested',
within a single dome:
living only a year, they set
much seed to ensure survival.
To admire their miniature beauty
you must use a pocket-glass,
kneel, and be thought eccentric
by those who regard the world as designed
to be viewed from five-foot-six.
Each flower is formed by lobes
dividing from a trumpet-tube:
the upper two are close-joined,
pricked up like rabbit's ears;
the other three widely spaced
and almost horizontal;
three stamens jut from the centre.

As you stand once more, your eye takes in afresh
the carpet of pink against the grey of the rocks
and the ever-changing darkness of the sea.

Stork's-bill, *Erodium cicutarium*

The pale-pink flowers appear amid fine-cut foliage
through spring and summer in places dried by the sun.
They are evanescent: I have watched the five petals drop
in the space of a minute, with no wind nudging them,
as if on command, discarded by the plant.
Other geraniums are easily told apart:
Herb-Robert has flowers in twos, and is strongly scented;
Dove's-foot (*Geranium molle*) has leaves that are lobed,
not finely cut, and its petals are notched at the tips.

Once Stork's-bill has dropped its petals, a drama begins.
The pistil stretches out to a startling length,
as long as the top two joints of your little finger,
slender and tapered, just like a sturdy needle
or the probing bill of a stork or crane or heron.
Around its base are packed five slender seed-pods,
each like the head of a spear, and each with a long
straight wiry tail where the haft of a spear would be.
When dried by the sun, the spear-head shoots from the plant
and plunges point-first down among the grasses.
What happens next is a wonder:
that long straight tail begins to twist like a corkscrew,
gimletting, driving the seed-head into the soil.
What a demonstration of nature's urgent force,
intent on impelling the seed to its proper bed!
You can watch the process closely: cut a ripe pistil,
leave it on a table in the sun, and go on with your work.
A seedhead will suddenly shoot itself over the table,
making you jump along with it;
and at once, in the warmth of the sun or the palm of your hand,
the tail will commence its coiling, augering movement.

Spring Gold, *Lomatium utriculatum*

In the chilly air of spring
press your hand to the turf
of a slope that faces the sun,
and feel the warmth trapped there
amongst the grasses and moss.
Here the Spring Gold nestles,
catching the heat of the sun
in nets of fine-cut leaves,
and echoing its colour in clusters
of golden-yellow flowers.
    Each of these clusters is an *umbel*:
its flowerstems rise from a common point,
like an *umbrella's* stays,
and the flowers are so close together
that they form a single surface,
which is intricately textured
by petals, stamens, sepals.
A plant will have perhaps ten
of these clusters, all of them rising
from a common point on the stem
to form a *compound umbel*,
a bright concretion of gold.
    As a prospector, toiling
in an icy mountain creek
will suddenly see the glimmer
of tiny golden nuggets
beside the gravel in his pan,
and his heart rejoices within him:
so the wayfarer, braced
against chilling airs from the sea,
is warmed to find amidst the grass
these nuggets of springtime gold.

Arbutus, Madrone, *Arbutus menziesii*

Icons of the coast, these trees are told
at once by the colours of trunk and limbs,
well known in the paintings of E.J. Hughes:
the bark is reddish-brown when mature,
but then peels off, revealing green
and yellow-orange, which reddens again.
An oversized member of the Heather family,
Arbutus has roots that are tough and tenacious,
clamping their hold on layers of rock,
while the trunk leans over the water, pushing
the green world out to its furthest limit.
Often the trunk is oddly twisted,
as if it were corkscrewed into the rocks.
And often a trunk or branch will snake
away from the base for thirty feet
just above ground, to reach the light
--a wonderful place for kids to clamber
over limbs that are smooth and brown and bare
like those of a sturdy suntanned mother.
Like the roots, the flowers are those of a Heather,
miniature urns; in April and May
they are massed in cream or green-white clusters,
set off by the glossy dark-green leaves.
Back from the shore, the trees grow taller;
sixty feet up in the canopy
you will see the floating clouds of flowers.

They love the challenge of rocky places,
the rainswept summits of hills and mountains.
In the Great Flood, so the story runs,
all of Saanich was drowned by the rising
seas, except the top of Mt Newton.
There the sole survivors took refuge,
and there, according to Chief Philip Paul,
an arbutus rooted on the very summit
provided a place to secure their canoe,
lest it be lost in the swirling floods.
Since that time the mountain is called
Place of Refuge in the Saanich tongue,
ŁÁUWELṈEW̱; and the people still
do not burn arbutus wood in their stoves,
mindful of the service the tree provided.

Field Chickweed, *Cerastium arvense*

It greets you beside the sea
or on grassy bluffs in the hills
with the bright faces of its flowers,
above the tangled mat
of grey-green leaves and stems.
What they lack in size, the flowers
often make up in number,
gladdening the rocks and hillsides.
A good recognition mark:
each of the five white petals
is deeply notched. Up close,
from the green centre of the flower,
honey-lines lead the eye out
into the petals' brightness,
so the flower appears to spring
                upward and apart.
In moister places the stems
are long and leggy and supple,
but in clifftop gravel
they are stiff and short,
so the flowers quiver in the breeze.
Adaptable, tough, this plant
ranges from the rocky shores
of Newfoundland in the east
to Haida Gwaii in the west,
and in distant Europe or Asia
it offers a cheering welcome
to the wanderer far from home.

Golden Paintbrush, *Castilleja levisecta*

Beauty is more intense when endangered:
the light of the sun seems lovelier at its setting
when night begins to close around it.
So this plant, nearing the end of its days,
glows with a golden-yellow brightness
on the coast that faces the westering sun.
Though once more widespread, it is now imperilled,
for it grows in grasslands close to the sea,
the very places where we build our mansions
or establish parks for romping with our dogs—
meaning no harm, but by sheer numbers,
by our burdensome unrelenting presence,
we obliterate all that is rare and lovely,
and leave our children an impoverished world.
And we cannot reprieve the Golden Paintbrush
by moving it out of the bulldozer's path,
for it depends, like its Paintbrush siblings,
on interconnection with other plants,
nourished in part by their roots, in addition
to the food it creates in its own green leaves,
so it weakens and dies once torn from its place.
It survives now only on a few small islands
and in fields untrampled as yet by humans,
a reminder of the Golden Age,
a living heir of its radiance,
still lingering here on the coast.

Scotch Broom, *Cytisus scoparius*

The Scotch Broom sweeps along the coast
scouring out native plants,
like the squat house sparrows
evicting the graceful swallows.
An ungainly plant,
so like the invaders who brought it,
tenacious, aggressive and 'successful'
if success is measured by dominance.
On August afternoons you will hear the gunshot
of its black pods firing out seed
to colonize yet more land.

And yet, and yet—
who could deplore the blaze of buttercup-yellow
transforming the bulldozed rocks of the Malahat highway?
Who could disdain the moulded form of the flower?
Pick one and study its structure.
The upright *banner* or *standard* is curved for strength,
with fine orange lines at the base;
the vertical *keel* is tightly folded
to contain the pistil and stamens,
and flanked by two gold *wings* for greater protection.
Press down on the keel, as a landing insect would do:
stamens pop up in a puff of pollen;
the pistil springs even more fiercely, coiling right round,
to ensure that it grazes the visiting bee.

Many wish that the Broom
would brush itself back to Europe,
or that some helpful witch
would ride it into the sunset.
But since it is here, its springtime

colour is a compensation,
even if for the rest of the year
we curse it and cut it and root it
out of our pastures and parks
and the wilder places that are left.

Beach Pea, *Lathyrus japonicus*

Never far from the reach of the sea,
anchored in shifting sand
by tough, far-reaching rhizomes,
Beach Pea is well adapted
to withstand the rigours of its place,
the drying wind and sun, the soaking spray.
Its leaves are firm and leathery;
and while most peas are pliant,
Beach Pea's stems are sturdy,
for there may be no plants nearby
on which to cling for support
against the sweep of the wind.
(A good distinguishing feature:
at the base of each row of leaves
you will find an odd-looking pair,
roughly triangular—or halberd-shaped
as the textbooks like to describe them
for those familiar with halberds.)
The flower-stems too stand firm.
Their flowers are glossy-smooth,
of legume form, like Broom's,
with *banners* of reddish-purple
contrasted with blue on the *wings*.

Each flowerbud is already shaped,
as it swells, like a miniature peapod;
and after the flower is fertile
it returns to the selfsame shape,
with its peas aligned in their pod.
'Beanpod' was a name the Greeks
applied to a kind of yacht
for its long and slender shape.

Conversely the Haida bestowed
the name 'Raven's canoe'
on this plant, for the shape of its pods.
Just like a Haida canoe,
the pods and peas can voyage
many miles on the shifting tides,
then beach on an unknown shore.
There the tough outer coat
of the peas is abraded by the scouring
of waves on the beach, so they sprout
and anchor themselves in the sand,
to grow within reach of the sea.

Thrift or Sea Pink, *Armeria maritima*

This frugal plant
is well named Thrift,
for it subsists
and even thrives
in the merest sift
of soil in a cleft,
a rift, a fissure,
a meagre crevice.
With its twiglike root
it grips the place
and taps for moisture.
You will find its tufts
of grasslike leaves
just feet above
the reach of the tide.
Amongst them rises
the six-inch stem,
bare, red-tinged,
wiry and tough,
sustaining a head
of clustered flowers.
Five brown sepals
clasp the head
to support its form;
and each tiny flower,
pink with purple,
is held contained
in a papery cell
to afford protection.
So flowers and stem
together resist
the restless wind,

the splashing spray;
firm yet limber,
not stiff and rigid,
they weather hardship,
bend without breaking.

Nodding Onion, *Allium cernuum*

You hardly notice the skinny leaves
since they look like blades of grass, but they soon
announce their presence if you happen to tread them
by their characteristic onion odour,
common to garlic, leeks and chives.
Many first nations steamed the bulbs
in cooking-pits for sweetness and softness.
Now, with Eden depleted, we cannot
be digging bulbs from the wild, but we may
still gather a portion of the plentiful seed,
sow it in a corner of Eden in our gardens,
and watch the beauty of the flowers unfold.
The scape that bears them is pliant at first,
floppy as the leaves; later it strengthens,
but still retains at the top a springy
bend, so the clustering flowers as they open
'nod' in the breeze. (This shepherd's crook
is unusual in the Allium genus: *cernuum*
is a rare Latin word for this rare feature,
meaning 'head-first', and so 'downward-facing'.)
The flowers are delicate bells,
projecting long and elegant stamens,
and in understated colours, ranging
from white to a rosy pink; for me
the finest is a white just washed with pink.
As they age, the scape at last pulls upwards,
holding aloft the tight round pods
so they can dry and scatter their seed.

Sedum, Stonecrop
(various species, especially *Sedum spathulifolium*)

The sedentary sedum
squats on broken rocks
and clings to the faces of cliffs.
From its stiff and pudgy leaves
with their dusty grey-green surface
you would think the plant absorbs
the stony substance of its place
—then magically transmutes it
as the leaves take on new colours,
some in winter, some in summer,
orange and deep rose-red.

The first hot suns effect
another transmutation,
to a carpet of brightest yellow,
as the flowerstalks open their buds
and hide the leaves from sight.
Examine an opened flower
and encounter a different image:
five slender, wide-spaced petals,
firm and shining, fashion
a delicate, formal star;
other stars cluster round it,
a close-knit constellation
just inches above the earth.

Now come the long dry days,
dog-days of summer, and now
a further marvel, for sedum
survives under scorching suns,
amidst rocks too hot to touch,

where moisture is long since dried
from the parched and hardened soil.
It lives on liquid hoarded
in those fleshy succulent leaves,
whose dusky, waxy coating
resists evaporation;
and so it endures the drought,
till autumn's rains arrive.

Yellow Sand Verbena, *Abronia latifolia*

From beaches along the coast
and sandy dunes there emerge
heavy creeping stems
and rubbery, rounded leaves.
Grains of sand stick to them,
especially on the underside:
you would think the plant was fashioned
by alchemy out of the beach,
and some of the raw material
clings to the finished product.
In fact the stems and leaves
have sticky glandular hairs,
that hold the chunks of sand,
helping to weigh the plant down
and anchor it against the wind
—which is magic quite as potent.

In June the stems curve upward
to carry rounded heads
of lemon-yellow flowers,
vivid, with flaring petals
and an exquisite scent.
Here again is magic:
Nature as alchemist,
transmuting barren sand
into such entrancing flowers
and a perfume so bewitching.

Gumweed, *Grindelia integrifolia*

When the soil is parched and dry
and other plants are brown,
Gumweed retains green leaves
and thrusts out yellow flowers
from June well into Fall.
Never far from the sea,
the leaves are narrow and tough
to withstand the drying winds
and desiccating suns.
Likewise the flowers are chunky,
carried on sturdy stems,
with shortish gummy petals.

Plants of the Grindel genus
endure in the harshest places,
arid or poorly-drained.
And they lend their strength to humans:
for the spiky flowerbuds exude
a sticky medicinal gum
which covers the buds so thickly
you would doubt they could force their way open.
The virtue of this gum for relief
of burns and ivy poisoning,
asthma, bronchitis and whooping-cough,
was long since known to first peoples
and taught by them to whites.
The name *Grindelia* honours
the Russian David Grindel,
learned in plants and cures.
We might honour also the lives
of countless traditional healers,
nameless, whose knowledge endured
and helped their people endure.

Labrador Tea, *Rhododendron groenlandicum* (formerly *Ledum gr.*)

This is a scraggly plant, whose stems
look twisted and bent with age. It grows
in sphagnum-spongy places, cold
acidic bogs. What you notice first
are the narrow, sickly-seeming leaves,
with crinkly surfaces, and edges curling
downwards, as though they were dying of drought
in the midst of all that oozing water.
Though evergreen, they quickly droop,
with brown appearing on the upper surface
and a red-brown fuzz beneath the leaves,
as if they were turning rusty.
                                        And yet
the plant does have one redeeming feature:
although the flowerbuds show no promise,
not plump and swelling, but wrinkled grapenuts,
they produce fine flowers in clear-white clusters,
well contrasted with the dark-green leaves.
Each flower is an open, five-petalled star.
What catches the eye is the fact that the stamens
are also white, and extend well clear
of the flower, so they seem to project its colour
outwards in a spray.
                              And yes, there is
one other redeeming feature: the leaves
when crushed are fragrant, and if they are steeped
in boiling water like breakfast tea,
they provide a pleasant aromatic drink.
Fine-tune the strength, and adjust the flavour
with lemon or sugar, according to taste.

The Haida relished it dark and strong,
infusing it for several days,
as British grandmas sketched by Giles,
fierce and steely and living to ninety,
toughened their stomachs with well-stewed tea.

Ocean Spray, *Holodiscus discolor*

Toughest of shrubs, the Ocean Spray
withstands the spindrift, the winds of winter,
the drought and suns of summer.
Yet its upright stems are tipped in June
with intricate airy sprays of flowers.
First you see clusters of tiny spheres,
formally spaced along their stalks,
like small-scale models of molecule structure.
But as the days grow warm
each sphere bursts open, a slow explosion
of cream-white petals and stamens,
and the clusters change into lacy structures,
splatters of sea-foam frozen in the air.

The wood of the stems is hard, and fire
will temper it to greater hardness.
Before iron came, and iron men,
it served to make tools and weapons:
digging-sticks for gathering bulbs;
shafts for arrows, spears, harpoons;
halibut hooks and fish clubs;
sticks to dry clams and barbeque salmon,
since the wood does not spoil the flavour;
needles for weaving cat-tail mats;
hoops for drums and for babies' cradles.
Settlers learnt its use from the natives,
and Lewis Clark tells of seeing a barn
fastened together with pegs of this wood,
with never a single nail.

So Ocean Spray was prized,
and latterly known as "Ironwood" for its strength.
But with all its toughness, Ironwood cannot stand
against the bulldozer's scouring blade,
as men of iron and steel
"develop"/destroy the land,
with no sense of place or limit.
Machines go into the thickets,
the ancient homes of birds, untouched for aeons.
The Douglas Firs, sky-reachers, are overthrown;
the Arbutus is felled, survivor of many a storm,
the white-flowered Dogwood, loveliest of trees,
and the ancient Oak that shades wildflowers in summer
and remembers the land before the white man came;
Redcurrant is smashed, Mock Orange and Ocean Spray;
Lilies, Satinflowers, Camas are crushed into dirt.
You return to the well-known place,
to find only broken stems and ravaged earth,
the air denuded of trees and branches and birds,
and empty of all but the restless winds from the sea.

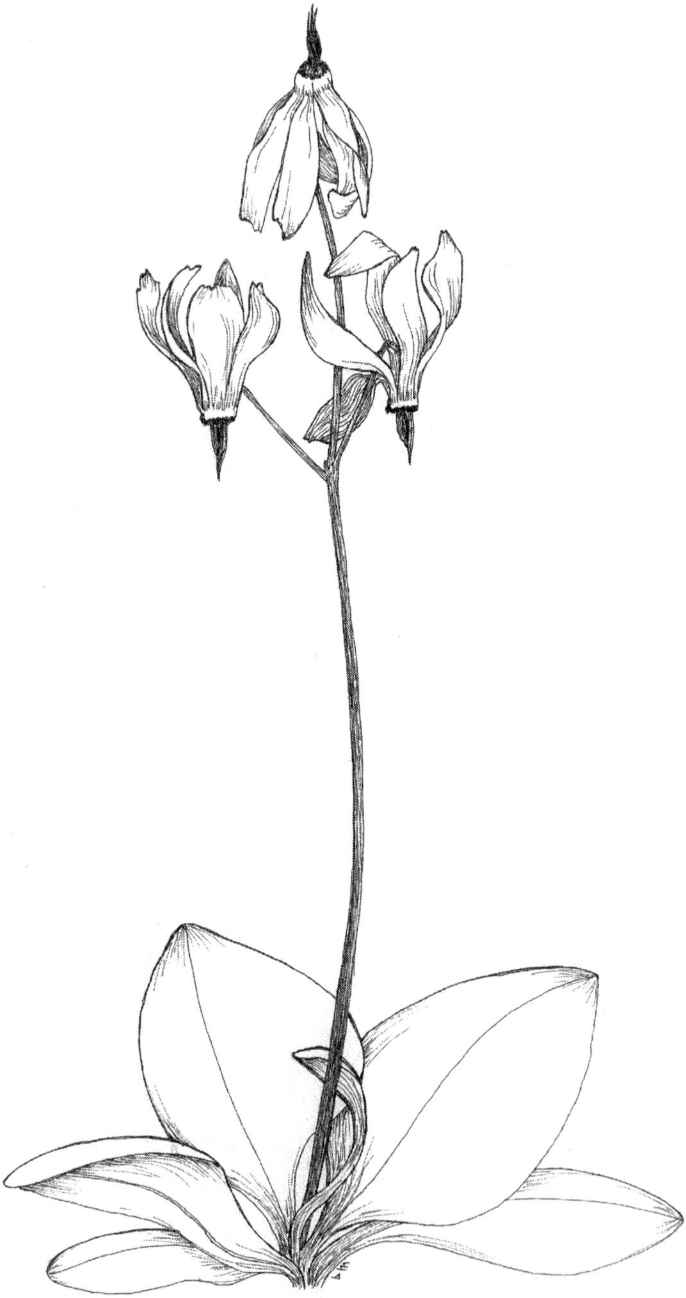

Shooting Star, *Dodecatheon hendersonii*

Part 2:
FIELD AND SLOPE

Shooting Star
(*Dodecatheon hendersonii* more common on the Island, *D. pulchellum* more common on the mainland)

Dóh-decathée-own: savour the Greek,
'of the twelve gods,' the pantheon.
A mysterious name, to decode as you will:
for Pliny, it commended a beautiful plant,
"linking it with the majesty of all the gods."

They grow in response to the strengthening light
and quickening showers of March and April.
At the base is a circle of fleshy leaves,
rounded and flat in the Island species,
erect and pointed in *D. pulchellum.*
Above them the sturdy scape supports
a flight of flowers suspended,
like feathered darts arrested in fall,
or brilliant hummingbirds poised in the air,
their long beaks tilted down to drink:
the clustered stamens come to a point,
sharp as a needle,
while the petals stream and flare behind them.

What rich variety of colours!
In a single patch you will find a range
from reddish-purple to paler orchid
to palest pink, and rarely white.
The Island species is sharp and classic,
with a whitish ring at the base of the petals
set off by stamens of purple-black;
pulchellum is warmer, with a deep-yellow ring
at the petals' base—'a diminutive beauty.'

This ring is an aid in pollination,
a guide to the visiting bumblebee:
she clasps it, which brings her body in contact
with the clustered stamens; then a buzz of her wings
shakes pollen onto her belly, to be stored
in sacs on her legs, but some to be brushed
on the next Shooting Star she finds on her rounds.
Once they are fertile, the flowers turn upward
and swell into pods, first green then brown,
no longer poised and attractive and chic,
but capsules carrying the seeds of life
to create fresh beauty in years to come.

Camas (shorter Common Camas, *Camassia quamash*; taller Great Camas, *C. leichtlinii*)

Meriwether Lewis beheld them blooming
in natural meadows in the glorious bounty
of the Golden Age, and he "could have swoarn"
he was seeing "lakes of fine clear water,"
so bright and unbroken the pools of blue.
Up close you can understand the deception:
each plant has many flowers on its scape,
and some are a light sky-blue, but most
are cool and deep as a mountain lake.
Their bright-yellow anthers set them off,
floating freely (their stalks of blue
almost unseen), and echoed in colour
by the buttercups glistening in the selfsame meadow.
The Common Camas flowers in April,
as skeins of geese are flying north;
Great Camas is later, but they overlap.
Tell them apart by the shape of the flower:
the smaller plant has a single petal
pointing down, while the other five
form the upper half of a circle;
in the taller camas, all six petals
are evenly spaced and make a full circle
—a receptive saucer tilted upward,
a satellite-dish for the bright blue sky.

The meadows around Victoria
were once abundant with them
under the open trees,
Arbutus and Garry Oak.
The native Salish people
dug the bulbs for food,
and traded them along the coast.
They would tend the grassy meadows,
removing shrubs and rocks,
and rooting out Death Camas,
Zygadenus venenosus,
which grows in the camas fields
as Nature's counterpoise.
At harvesting-time the Salish
would pry the sod up with sticks,
remove the larger bulbs,
and then reseat the turves,
so that the smaller bulbs
could grow for future harvests.
Gathering the bulbs in thousands
they would steam them slowly, gently,
in covered pits, between layers
of bracken, salal, Grand Fir,
for twenty-four hours or more,
then feast on their sweet, substantial
flesh with fish or meat,
drying others for trade or storage.
        For the Saanich people, the islets
close to the coast were a favoured
site for collecting the bulbs:
the soil is shallow but rich,
made fertile by breeding gulls.

While camping here they would feast
on the rich supplies of gulls' eggs,
each one larger than a chicken's.
But they took good care to lift
only one or two from the nest,
lest the mother abandon the clutch,
and the cycle of life be broken.
        So they lived in balance with Nature,
knowing that all her bounty
depends on moderate use.
But now the pattern is broken,
the natives dispossessed,
the land not used, but exploited
in ways it cannot sustain.
And in Oregon, 1876,
the Nez Percé, who fought
to retain their gathering-places,
were slaughtered by the U.S. Army,
so that the settlers' pigs
could root out the camas bulbs.

Red Elderberry, *Sambucus racemosa*

'Elder' has nothing to do with age.
This shrub is sappy and full of vigour,
among the first to unfurl its leaves
in spring on the coast, and later displaying
creamy clusters like rounded cones
filled with flowerlets. A cluster of this shape,
well known in lilacs, can be called a *thyrsus*,
named for the cone-shaped tip of the staff
of Bacchus—god of all fluid life,
"the liquid fire in the grape, the sap
thrusting in a tree, the blood pounding
in the veins of an animal, all the tides
that ebb and flow in the life of nature,"
flowing also in the stems of the youthful Elder.
As a god of the wild, Bacchus loved the music
of whittled pipes, such as those of Elder.
Its stems are pithy and easily hollowed,
and were latterly used round the Salish Sea
for whistles, blowguns and pipes for smoking.
But the stems are toxic when fresh: as always,
know your material, or ask those who know.

Enough digression: back to the flowers!
When first they open they are sweetly scented,
but soon grow tangy, to attract a host
of insects toting pollen; hence,
in summer, masses of bright-red berries,
eagerly sought by Band-tailed Pigeons,
grouse and bears and other creatures.
Some proscribe them as poison to humans,
but this is as foolish as eating them rashly.

Indeed the berries can be mildly toxic
if eaten unripe, inducing nausea,
but when ripe they are fine. The safest course
is to cook them first, and produce, for example,
a jelly, with sugar added to taste
and lemon, or perhaps other native berries.
So careful experience of particular plants
in particular places builds up the store
of local knowledge, to be valued as highly
as the gifts of nature themselves are valued.

Oregon Grape, Mahonia
(taller species, *Mahonia aquifolium*; shorter, *M. nervosa*)

Look for the leaves, dark green and prickly,
like those of holly. Its prickles are 'friendly',
turned sideways rather than outwards, but still
they will sting your hand if you clutch the dry leaves.
Aquifolium would be better spelt
acufolium: it means 'prickly-leaf'.
The taller species loves sunny places,
and its leaves are so glossy that they mirror the sky;
the shorter form loves the shade of firs,
and its leaves are duller, matching its home.

Both plants carry trusses of yellow flowers
clustered together, worth a close look.
The buds are tiny balls of yellow.
As they open, they keep their spherical shape,
with an inner circle of curving petals
and an outer of short and sturdy sepals,
like segments of orange-peel joined at the base.
After they fall, the rounded form reappears
as the fruit swells out into spheres of blue,
dusky with 'bloom' like plums and apples.

The berries are tart, but make a fine jelly
especially mixed with salal (see Pojar!).
And the plants offer other uses too
for ingenious humans. The inner bark
is a startling, vivid orange-yellow,
as you find if you scrape or snap a twig.
It contains an alkaloid, berberine,
useful in medicine; and for early peoples
it provided a bright-yellow dye that would fix
without mordants, to colour materials for baskets.

But the commonest use is by gardeners, who grow
the taller species as a hedge or backdrop,
and prize it for its evergreen lustrous leaves,
still showy in winter. When new, the leaves
are the colour of copper; then they darken to green;
as a final glory they have sunset colours,
the orange of flame, a deep vermilion,
a barbaric scarlet.

Strawberry, *Fragaria* (various species)

Bright flowers turned up to the sky,
with five broad petals of the purest
white overlapping each other,
and centres yellow as the sun.
Once fertile, the flowers redden
as if flushed with warmth and love
and anticipating the redness
of the ripened sun-warm berries.
Bella Coola children would search
in the drowsy days of July,
crying out *kwululuu-tsuu-tsu*,
"I've found another strawberry",
whenever they found a patch
or even a single berry.
And at three years old my daughter
walked on, though tired, across Strawberry Flats,
recharged at every few steps
by a tiny droplet of sweetness.
Wild Strawberries! Bergman's title
was vivid for a city boy;
it evoked a pristine world
where nature's gifts could exist
untrampled, unpoisoned, unbulldozed.
Ovid recounts that *fraga*
along with cornels and cherries
and the fruit of the Arbute tree
were foods of the Golden Age
—an age when the earth gave freely,
unwounded by the teeth of the plough,
when spring continued forever,
and time, unthought-of, stood still.

And for all whose childhood was happy
with uncles larger than life
and tree-forts and strawberries and beachfires
and swimming and sunny days
unending, that age still lingers
in the golden fields of the mind.

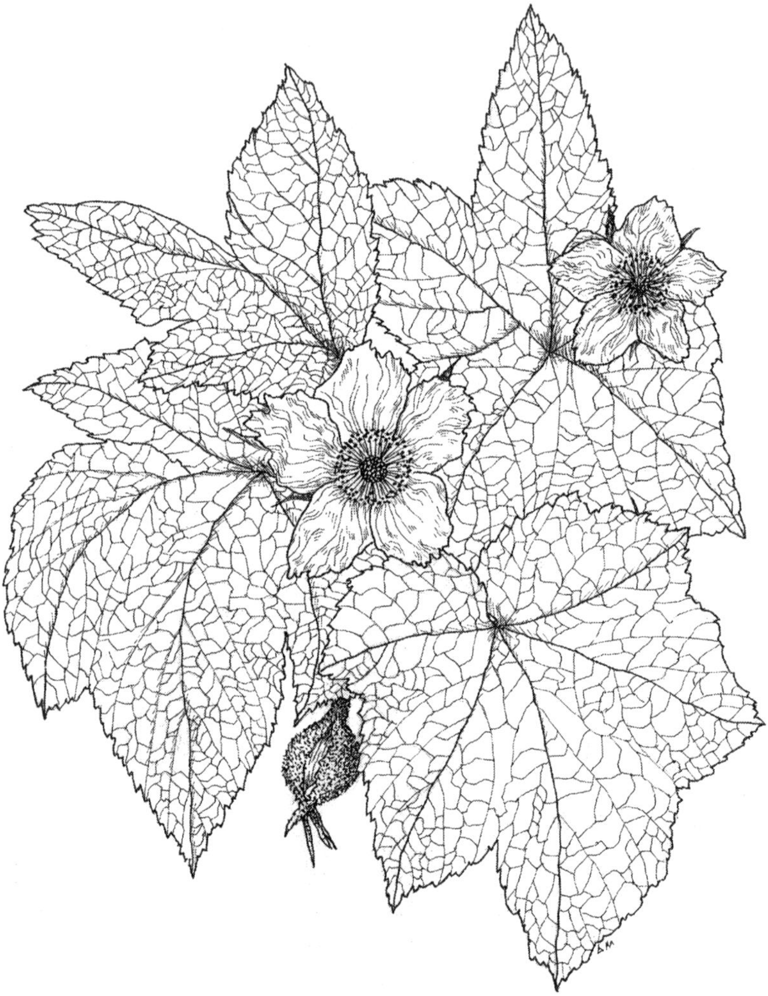

Thimbleberry, *Rubus parviflorus*

Some thimble! The berry is more
like a shallow dome, a skull-cap.
Even on the underside,
where the berry was held on its white
receptacle, the hollow
is more of a cone than a thimble.
The flesh is soft as a raspberry,
needing only a nudge to fall ripely
into your hand, or your mouth.
The flower is fine and showy,
like a Nootka Rose (but smaller,
and white not pink), with five
broad overlapping petals,
crinkled like tissue paper,
and with shaggy stamens at the centre.
The leaf: light-green, five-pointed,
with the shape of a maple-leaf
but soft as velvet, and crinkled
just as the petals are crinkled
—a gentle container for the squishy berries
if you fasten the points together
to make an impromptu cup.
The canes: erect and smooth,
rarely above head-height.
Contrast the Salmonberry,
which is taller, twiggier, pricklier.
A stand of Thimbleberry
grows dense but trim and handsome,
neat, like a natural hedge.

First peoples would pluck the shoots
from the canes in early spring,
peel them, and eat them raw,
finding them sweet and juicy;
Nuu-chah-nulth canoes could be seen
laden with these new shoots.
The berries can be eaten fresh,
but are seedy and sometimes insipid.
Their perfect use was for drying
into cakes, with other berries
or in layers with roasted clams,
pressed under weights, and placed
on a mat of the gentle leaves.

Pacific Crabapple, *Malus fusca or Pyrus fusca*

You will find them in cool damp soil
at the edges of streams and lakes,
with the spreading growth of the fruit-tree kind,
and mingled clusters of flowers and leaves
at hand's-lengths along the stems.
They will dazzle you in the Maytime sun
with their glossy new leaves, even lighter
when turned by the wind, and the waxy
whiteness of the handsome flowers.
Such flowers! Well-bred, sweet-scented
as on any orchard tree:
their petals broad and rounded,
like those of the strawberry flower,
and cupped, forming shallow dishes
—fertile, receptive shapes.
Next come the egg-shaped apples,
small but plentiful, hanging
in long-stemmed clusters, and mellowing
to yellows and reds in the fall.

First peoples would pick them in abundance
while green, to ripen in storage;
being acid, they were easily kept,
and grew softer and sweeter with time.
To have many bentwood boxes
fragrant with cedar and the scent of apples
was a sign of wealth and high standing.
Connoisseurs of these apples, the Gitga'at
of Hartley Bay distinguished
and named six separate varieties
by shape and colour and taste.

On the Island, the Nootka people,
as Gilbert Sproat recorded,
were "as careful of their crab-apples"
at the head of Alberni Canal
"as we are of our orchards;"
but as white settlers encroached on their gathering-places,
the Nootka, in anger and despair, at fruiting-time
cut down the trees around the settlers' farms
so that their last crop of fruit
should at least be easy to harvest
as the trees lay on the ground.

Oxeye Daisy, *Leucanthemum vulgare* (*aka Chrysanthemum leucanthemum*)

Millions of Oxeye Daisies are blooming
in June as the children tumble out of school,
and summer is on us. The plants arrived
with the early settlers, and spread to the west
like a silent storm; and now they whiten
fields and hillsides with billowing drifts,
the snows of summer. The big round centres
of the flowers, with their fringe of long firm petals,
recall the great fringed eye of an ox—
high praise for beauty, since Homer speaks
of elegant ox-eyed women of Troy
and of ox-eyed Hera, wife of Zeus.
Another image that the golden centre
suggests in summer is the eye of day,
days-eye, the sun, whose rays are the petals.

As in several daisy species, the 'flower'
is composed of many miniature flowers,
florets: the 'petals' are *ray-florets*,
while the disk at the centre consists of hundreds
of *disk-florets* in a golden cluster.
Up close, it looks like a beaded button,
in which there are intricate swirling patterns
formed by the rings of tiny buds.
These minute flowers begin to open
at the fringe of the button first,
so a visiting bee will walk a circuit
around the edge; only later,
as summer matures, does she reach the centre.

Chocolate Lily, *Fritillaria affinis* (formerly *F. lanceolata*)

'Chocolate' sounds enticing,
and Briony sinfully invites you
to fritter away a decadent afternoon
among the fritillary lilies.
If you come, don't choose a dull day,
for then these lilies are drab
and elusive: among the brush
and the grasses, the bell-shaped flowers
are camouflaged by their colours
of greenish-yellow, mottled
with chocolate (milk or dark).
So come in the sunshine, and see
how the sunlight softly blends
with the lustrous sheen of the petals,
as in Lewis Clark's sensuous photo.
Then gently raise a flower,
and watch how the light shines through
the yellow parts of the petals,
and glistens inside the bell.
Rarely you will find the flowers of a plant
completely lacking the chocolate pigment—
bells of pure yellow, a wonder.
        Plan too on a later visit
in summer, to admire the sturdy
capsule containing the seeds.
It is bell-pepper-shaped (but smaller!),
with each of its vertical sections,
six in all, projecting
into a knife-edged ridge.
You would think it a man-made object,
almost unreal, like the flower.

Can Chocolate Lilies be eaten?
The starchy bulbs are covered
with nodules like grains of cooked rice.
They were formerly tended and gathered
and cooked and eaten, like camas.
But we schlubs have made the plant scarce
by destroying the meadows where it grew,
so it needs to be spared; it serves
no longer as food, but still
as a chocolate treat for the eyes.

Mock Orange, *Philadelphus lewisii*

In June the edges of fields and forests
are fragrant with scent like orange blossom,
from flowers well matched to the month of marriage:
white blossoms decking the white-veiled bride
betoken purity and fruitfulness,
opposites that go hand in hand together.
The flowers of Mock Orange appear in groups,
never alone: they are formed by four
cupped petals, smooth, of pure milk-white,
set off by golden-yellow stamens
and light green leaves that grow in pairs.
Both leaves and flowers, when bruised and rubbed
robustly in water, produce a lather
with cleansing properties, to bathe the skin.
The wood of the stems is hard, a fine
resource for people who knew their plants.
Sam Mitchell described to Nancy Turner
how he fashioned knitting-needles for his wife
from an inch-thick stem: he split it in quarters,
whittled the pieces, hardened their points
with fire, and sanded and polished the surface—
a practical labour of married love.
I think of Mock Orange as smiling on union,
on harmony, on all social concord
(*Philadelphus* refers to brotherly love),
and as serving the eternal goddess Juno,
patron of marriage, helper in childbirth,
whose spirit is active still in her month of June.

Red Columbine, *Aquilegia formosa*

How can this complex flower
with such an intricate structure
be caught in a web of words?
I should make an attempt, at least,
beginning with the stalk, which curves
down like a shepherd's crook.
Arranged around it in a circle
are five inverted trumpets,
coral-red on their stems,
yellow at the open mouths.
Between their stems five sepals,
also coral-red, flare out;
and underneath the flower,
where the trumpet-mouths are joined,
there hangs a bundle of stamens
bearing their yellow anthers,
an upended harvest sheaf.
—A description of sorts, but who
could express the harmony of the whole,
or the way that the sepals carry
the red of the trumpets outwards
so they seem to be part of their form,
though really a separate structure?
          Look again at those trumpets, and notice
how the narrow end is capped
by a swollen bead, a *spur*
filled with honeyed nectar.
Hummingbirds thrust their beaks
deep into the trumpet to reach
this treasure, and as they hover
they brush the pollen from the anthers
onto their beating wings.

Someone imagined this flower
as a group of five elegant doves
gathered in a circle to drink,
and named it Columbine
from Latin *columba*, a dove.
The down-curved stalk suggests
the spring from which they sip;
the spurs are their inclined heads,
the stems, their swelling bodies,
and the flaring sepals perhaps
represent half-opened wings.
Now step back, and admire
the airy grace of the plant,
its delicate light-green leaves
surmounted by wiry stalks
which position the flowers just so,
at a proper distance from each other.
This is an effect that depends
not on individual beauty,
but on *ma*, on intervals,
on relationships in space:
the person who could best respond
with human skills would be
a master of Japanese brushwork,
who would have such knowledge in his fingers,
without any need of words.

Menzies' Larkspur, *Delphinium menziesii*

As the slopes begin to dry,
they are brightened by sturdy spires
of vivid, electric colours:
violet, purple-blue,
azure, pastel blue,
and rarely white, still lightly
pencilled with lines of blue.
Examine the elegant form of the flower:
the five broad 'petals' are really sepals,
while the true flower sits demure at the centre;
it has two white ruffs, like a starchy blouse,
and below them a pair of genuine petals
with wavy edges, falling like skirts,
a modest cover to hide the stamens.
But the nectary holds the greatest marvel.
The topmost sepal curves at the back
into a long cylindrical *spur*,
which sheathes a second tube, extending
behind those two white starchy ruffs,
and secreting honey to be found by long-tongued
epicures, butterflies and bees.
The heavy bumblebee lands,
parts the skirts by her weight,
and thrusts her tongue to the honey,
both taking and leaving pollen.

Delphinium is a metaphor:
the Greeks, perennial sailors,
named these flowers for the dolphin,
with the drooping end of the spur
representing the dolphin's nose.

In the flowerbud *before* it opens
you can find another resemblance:
the honey-tube now becomes
the dolphin's tail, while the sleek
and streamlined bud is his body;
the hollow at the front, where the flower
will open, is his open mouth.
So many likenesses! 'Spur'
itself is a metaphor,
elaborated in Europe:
in England the plant became Larkspur
from the spur on the heel of the bird;
in Germany, *Rittersporn*,
"the spur of a mounted knight."
Metaphors in plant names:
Satinflower, Stork's-Bill, Salmonberry,
and Columbine from doves.
This is how we humans think:
we delight in the act of finding
similarities, connections,
that make the world familiar
and enable us to name it.

Woolly Sunflower, Oregon Sunshine, *Eriophyllum lanatum*

Gold is best among metals;
yellow is best among colours,
shade of the life-giving sun.
These golden-yellow flowers
stretch towards the sky,
their element of light,
from rocky hillside ledges.
They lack the extravagant height
of the giant annual sunflower
(which is no relation); this plant
grows in a compact cushion,
never above knee-height.
The botanical name belabours
the downiness of the leaves,
with 'wool' in Greek *and* Latin
(Greek *erion*, Latin *lana*)—
'the woolly woolly-leaf.'
Better is the common name,
Woolly Sunflower, for that
is a mixture, an oxymoron,
a juxtaposition of contrasts,
and so is the plant itself:
its leaves grey-green and fuzzy,
the better to hold in moisture
in the drought of its sun-drenched home,
but the flowers sharp-cut and vibrant.
Their brightness will dazzle your eyes
if you look at them close in the sunlight,
as if they glowed with a portion
of that blazing primal fire.

Nootka Rose, *Rosa nutkana*

It thrives in open places,
roadsides, abandoned fields,
the edges of beaver ponds.
Its stock, deep-rooted and sturdy,
produces abundant flowers
borne singly at the tips of the canes,
which are armoured with serious thorns.
The flowers are simple but showy,
often two or three inches across,
their wildness emphasised
by the unkempt stamens at the centre;
delicate and short-lived,
but rich in beauty and scent.
Their colours vary from plant to plant,
and even across one flower,
darkening towards the centre;
best is a deep rose-red,
but the palest of pinks, near white,
has its own old-fashioned beauty.
Then come the rounded hips,
bright-red and glossy, and matched
by the wine-red colour of their stems.

Roses have often attracted meanings:
they are emblems of Lancaster, York and England.
The Nootka Rose can hardly bear
such weight, being evanescent and artless;
just so, it serves as a perfect symbol
for the bloom of youth and beauty, for their
impermanence, for the passing moment.
Rosa nutkana: the label was assigned
because it was first collected for science

at Nootka Sound. But the name evokes
that fateful initial contact at Nootka
of English and Mowachaht, foreign and native—
archê kakôn, the beginning of evils.
Inscribed in the name are so many sorrows
for peoples of the coast: scourging disease,
abuse and destruction of resources, loss
of land, of language, culture, skills,
despair in reserves or in alien towns.
Within a century, those many villages
where people had lived for generations
on generations in tune with nature,
understanding the place, its wealth and its limits,
now lay abandoned, inhabited only
by squirrel and wren, salal and the Nootka Rose.

Tapertip Onion, Hooker's Onion, *Allium acuminatum*

Its flowering has such perfection
you would never believe it an onion,
unless you can find the long leaf
(they are mostly withered by now)
and rub it between your fingers
to detect the telltale odour.
Once the bulbs were dug
for food, but their tiny size
is a poor return these days
for the loss of the lovely flowers:
grow spring or Spanish onions
in your garden, or buy them at the store,
and leave the native bulbs
to feast our eyes with their flowers.

Each flower has a formal beauty.
Its colour is a royal rose-purple,
dusky in duller weather
but luminous in bright sun.
The petals are stiff, like parchment,
with sharp recurving points,
whence the name Tapertip Onion
(*acumen* means 'tip' or 'sharpness').
They come in two sets of three,
an inner and an outer ring,
precisely offset from each other.
A glass will distinguish their forms:
the outer three are crisp-cut,
elliptical, pure in colour;
the inner ones crinkle-edged,
more angular, and tending
to whiteness toward the base.

As they age, their colour drains from them,
leaving them perfect in form
but empty of all their glory.
But if you cut a stalk
as the flowers first open, and set it
in water, out of the sun,
it will hold its colour for months.
Remember, though, to cut only
those you have grown yourself,
with patient nurturing, from seed,
and allow the wildling flowers
to ripen and replenish their kind.

Harvest Lily, *Brodiaea coronaria*

The hay is cut, the swallows fledge their young,
strawberries are ripe, and June gives way to summer;
the rocky slopes dry out, and the air is scented
with grasses and fir-needles yielding up their oils.
Among those fallen needles and withered grasses
you will find the fresh blue stars of the Harvest Lily,
borne in a 'crown' (*corona*) of four or five,
even up to ten, on a single wiry stem.
Six petals, perfectly spaced, make up the star
in a subtle alternation, broader and narrower,
each petal with a darker stripe along its axis.
Their lustrous colours of violet, blue and purple
contrast deliciously with the light-filled centre,
where three white *staminodes* stand erect, to form
a protective fence around the yellow stamens.
How does the scape arise from the dry cracked soil
when the plant's own leaves have withered? By virtue of the bulb
deep-seated in the ground, and rich in nutrients
that enable the flowers to brave the drought, and set
abundant seed. In western Washington State
these plants can be seen by the thousands on grassy slopes;
on the Island it is much less common, and seen with delight
on a summer walk, like a glimpse of water in the desert.

Cow Parsnip, *Heracleum lanatum*

Herácleum is named for Héracles,
a mighty plant in honour of a mighty hero.
Like him, it ranges far, from seagirt places
to the meadows of mountain heights. Its leaves come early
and declare the presence of the plant; they are huge, like rhubarb,
a little uncanny, as if they had gone berserk,
as the hero was wont to do. From among them the scapes
can grow well over the height of an ordinary human,
and the *compound umbels* of whitish flowers that top them
can spread as wide as the span of your outstretched hand.

These plants are eagerly browsed by elk and bear,
yet not wiped out, since the animals wander widely
and in moderate numbers; but where sheep and cattle are pastured,
they quickly destroy the plants through overgrazing.
First peoples used the hollow stems for flutes,
blowguns, whistles and drinking straws, like Elder.
But most they used the fresh young stalks for food,
as springtime greens. They knew the importance of peeling
the stalks before eating, like celery or rhubarb,
since the outer layer contains a chemical
that blisters your skin; they knew the special places
where the tastiest shoots were found; they had the good sense
not to over-exploit the crop, unlike Europeans;
and they were skilled at distinguishing this plant
from the deadly poisonous hemlocks, *Conium* and *Cicuta*,
which are look-alikes for those lacking expertise.
And so they secured for themselves an abundant harvest
by understanding their place, the effects of their actions,
the qualities of the plants and creatures around them.

This was the golden life before the Fall—
bulbs and roots in the soil, deer in the forests,
berries and fruits in plenty on the bushes and trees,
clams on the beaches, salmon in the streams and rivers—
before men came to appropriate the earth,
to divide it up into squares and parcels, to plunder
its bounty with ruthless greed, to domineer
with ever-increasing numbers, and so at last
to toil to obtain what once was freely given.
Perhaps in the cycles of time it will come again,
such a golden age, if our greed is ever sated,
if we ever outgrow our childish discontent,
our petulance. The people of that age will be
not a loftier race, as Symonds prophesied,
but calmer and wiser, chastened by our mistakes;
not aiming to "plant man's lordship firm on earth"
or create and manage a paradise planned by committees,
but rather to allow a paradise to grow,
to recreate itself. They will be a people
who respect the presence of others, humans, whales,
flowers and lichens, trees and wolves and owls,
and who find sufficiency to be ample riches.

Tiger Lily, *Lilium columbianum*

Part 3:
FOREST

Indian Plum, *Oemleria* (formerly *Osmaronia*) *cerasiformis*

In the weak light of late winter
its shimmer of green dawns on the eye
among the browns and greys of the forest.
Come close: you find the leaves in groups,
jauntily flaring upward,
vivid and half-transparent.
The flowers spill out from among the leaves
in hanging clusters like open catkins.
They conjure a summer insect,
with the clean-cut leaves as sturdy wings
and the drooping flowers as dangling legs.
The flowers are powerfully scented,
also the leaves when crushed,
whence *Osmaronia*, 'the pungent plant.'
The white tones of the petals
are muted by the green sepals,
which blend in turn with the greener leaves.
From a distance the shrubs appear
like washes of palest green,
hazes of colour, lightest of tones in the forest.
Pioneers among plants, forerunners of spring,
they endure the drizzle and chilling showers
and grey half-light of the waking year.

Skunk Cabbage / Swamp Lantern / Yellow Arum Lily
(*Lysichiton americanum*)

A plant of fire
that chooses to grow
in the wettest soil,
the ooze of the marsh—
flames amidst water.

The stalk emerges from the mud
swathed in a yellow garment,
a *spathe* or specialised leaf;
then the plant 'loosens its tunic'
(the meaning of *Lysichiton*)
to reveal the nubby flowers
in patterned rows on the stalk.
'Lantern' pictures it well:
the stalk like a textured mantle
burning the gas of the marsh,
and the spathe as a hood, reflecting
the light from its shiny surface.
It flares like *ignis fatuus*,
that other lantern of the marsh.
Later the glow is obscured
by the rank, cabbagy leaves
that swell to a metre or more,
bloated by water and muck.
Just as rank is the odour,
of skunk or twice-boiled cabbage,
attracting gross pollinators,
blowflies and carrion beetles.

Bears emerging in springtime
sate their hunger on this plant.
But if humans taste it, the crystals
that exist in the roots and leaves
burn and sting their mouths,
hot, like ginger or pepper,
and sharp, like tiny needles.
And yet first peoples could eat
the stalks of the leaves and the roots
to relieve late-winter famine.
How? By thorough roasting
or boiling, which breaks the crystals
or orders them, leaving them harmless:
so fire can be quenched by fire.
          And Ditidaht healers would use
the leaves as poultices,
laid underside down, to dress
the severest burns, where the skin
was gone: the wounds of fire
were eased by the fiery plant.

Salmonberry, *Rubus spectabilis*

Once Salmonberry grew
only in a single patch,
a sacred spot, well guarded.
This challenged Raven to exert
his cunning to steal the berries.
He succeeded, but as he flew off
he laughed so hard at his hoax
that the berries spilled out of his beak
all over the land below.
So a Chilcotin legend
explains the plant's dispersal,
and warns against losing a prize
through being too pleased at your triumph.

You will find its clusters of canes
armoured with short straight prickles
on the banks of streams and marshes;
at head-height or higher still
          rose-red flowers spaced out
                    along the branching twigs
and 'showy' against the darkness
and drabness of lingering winter.
The petals are crinkly and soft
like crêpe paper, yet each
sturdily curving outward.
And never far off is the buzz
of hummingbirds seeking their nectar
when little else is flowering.

Coastal peoples would peel
and eat the fresh green shoots
with salmon or salmon-roe.

The berries are raspberry-shaped,
some sweet, some tasteless and seedy;
purple and red and golden--
the last, with their beaded surface,
resembling salmon-spawn
and making a visual connection
with the salmon of the streams below.
Indeed the Nuu-Chah-Nulth people
knew their ripening as a sign
to watch for the sockeye salmon
at the start of their spawning migration.

What we label Swainson's Thrush
was known in many tongues
as the salmonberry-bird
(in Ditiďaht, *qaqawaši-yk*,
'salmonberry-making tendency');
for it nests at the margins of forests,
where thickets of salmonberry,
salal, and other shrubs
provide a supply of fruit:
so people gathering the berries
would notice the thrush, and hear
its tuneful ascending song.

Raven was once a guest
of this thrush, who sent out her kids
to gather berries for supper.
(So runs a Saanich story,
echoed up and down the coast.)
As they searched, she sang her lovely
song, and soon they returned,

their baskets loaded with ripened
berries fit for a feast.
"This is easy," thought Raven. Next day
he invited the thrush to *his* house,
and sent out *his* kids; as they searched,
he croaked out *his* song, *gaa gaaa*,
and they returned empty-handed.
So once again Raven the Trickster
got his come-uppance, and found
that some tricks are less easy than appears.

Yellow Wood Violet, *Viola glabella*

They struggle up from the dampness
and debris of the forest floor.
Their delicate heart-shaped leaves,
'smooth' and pointed, stretch
to catch the filtered light.
(Distinguish the Evergreen Violet,
with rounded, leathery leaves.)
Yet despite the gloom of their home
the yellow flowers are bright
like cheerful, open faces,
with the pencilled lines on their petals
like the crowsfeet of a smile.
These lines guide insects in
to pollinate the flowers
while fumbling to find the nectar.
But since this method is chancy
in flowers so close to the ground,
violets develop a second
series of 'flowers', without petals,
without scent, that never open,
but fertilize themselves
(*cleistogamous*, 'breeding shut-in')
to produce the life-bearing seed.

Red-flowering Currant, *Ribes sanguineum*

Its colour seems to emerge to the eye
from among the trees, the dappled sunshine
and the yellow-green of its own new leaves.
Rarely a pearl-white form is seen,
a marvel; more often, a blush-pink form,
especially in shade, within the woods;
and from pink a spectrum to rosy red
to strongest carmine--but 'blood-red' gives
too violent a picture for this well-bred plant.
Hummingbirds dart among the stems,
showier in their shimmering reds and greens,
but not more lovely than the still plant.

Each flower, as it opens along the cluster,
seems to delight in its own perfection.
If Indian Plum is blooming nearby,
you will see a resemblance: five-petalled flowers
of similar size, a fingernail wide,
in falling clusters ...
But an academic distinction: in the currant
the true petals form the central trumpet,
while the 'petals' around it, of deeper colour,
are sepals, forming the calyx or cup.

This plant has travelled far from its home.
The seeds collected by David Douglas
and taken to England, 1827,
more than paid for his expedition,
so highly was the beauty of the shrubs regarded.
Established as a garden favourite in Europe,
its fame travelled back, and now it is sold
in cultivar forms in garden centres

with double-white flowers, or crimson-red.
So the native forms, the country cousins,
may be left undisturbed in the woods and glades.
What a pleasure it is to encounter them
in their own true place, as their genial colour
gently, quietly announces their presence.

Orange Honeysuckle, *Lonicera ciliosa*

Fiery bursts of colour,
yellow and orange and red,
amongst the forest greens:
clusters of tube-shaped flowers,
each a blazing trumpet
accented by anthers that glow
flame-red, then fade to yellow.
Clasping each group of flowers
are two special leaves, combined
to form a disc—as it were
a sconce to contain the flames.
In Fall this sconce sets off
bright red translucent berries,
the fires of spring still glowing
in the dark autumnal woods.

The flowers are adapted to be tapped
by Hummingbirds, not by moths.
So they have no need, alas,
of the fragrance of their English cousins,
for the Rufous Hummers are drawn
by the brilliant colour, their own.
Only a Hummingbird's beak,
long and needle-thin,
could slip through the lengthy stamens
to suckle at the nectary;
but you can watch bees bore through
the wall of the trumpet at the base
and plunder the hoard of honey.

For all its sweetness, this vine
will kill young forest trees
—not with fire, despite its redness,
but by slower, stealthier means.
Its tough and wiry stems
clamber twenty feet up,
then coil around the 'host',
whether fir or dogwood, so tightly
that they cut right into the trunk.
Then at last the weight of wind
and snow will snap the tree
at the point that the vine has weakened.

Pink Fawn Lily, *Erythronium revolutum*

Sister to the Easter Lily, it displays the same
mottled leaves, and its flowers show off the same
gold anthers, the same orange flames at the centre,
the same tensed petals, but in this plant infused
with a rosy pink, and with the special grace
of rare and precious things.
It grows in a few secluded valley-bottoms,
where the soil washed down by creeks is moist and sandy.
Come in mid-April to the flats of Sutton Creek;
tread gently here, and wonder
at the host of sturdy stems like shepherds' crooks,
and the haze of colour from a myriad rose-pink flowers.
As you walk the trails, take in the forms of the place:
long-fallen logs with a blanket of moss and ferns,
and the grey-white verticals of living trunks.
Alders display their shimmering new leaves
and maples their hanging panicles of flowers;
the lower limbs, dying and laden with moss,
are soft and downward bent like gnarled old fingers.
Light falls through the limbs of trees to the forest floor
and is caught by leaves of fawn-dappled lilies, violets,
Vanilla-leaf, Bleeding Heart, False Lily-of-the-Valley.
Here and there are the sprawling fronds of Sword-fern;
the fiddleheads of the newer fronds, curled over,
echo the curving forms of the pink-lily stems.
If you see the pink of Bleeding Heart near the lilies,
observe the contrast: the lily more vibrant, less dusky,
of purer form, beautiful rather than pretty.

The peace of the place
invites you to walk
in love and reverence,
aware of the interplay
of plant and soil,
of light and shadow.
Yet people allow their dogs to trample the plants,
others ride trail-bikes and smash their beer-bottles here,
and logging-trucks roar past, belching their fumes:
Nature as humans' playground and backyard,
their stomping-ground, their dump, their source of plunder.

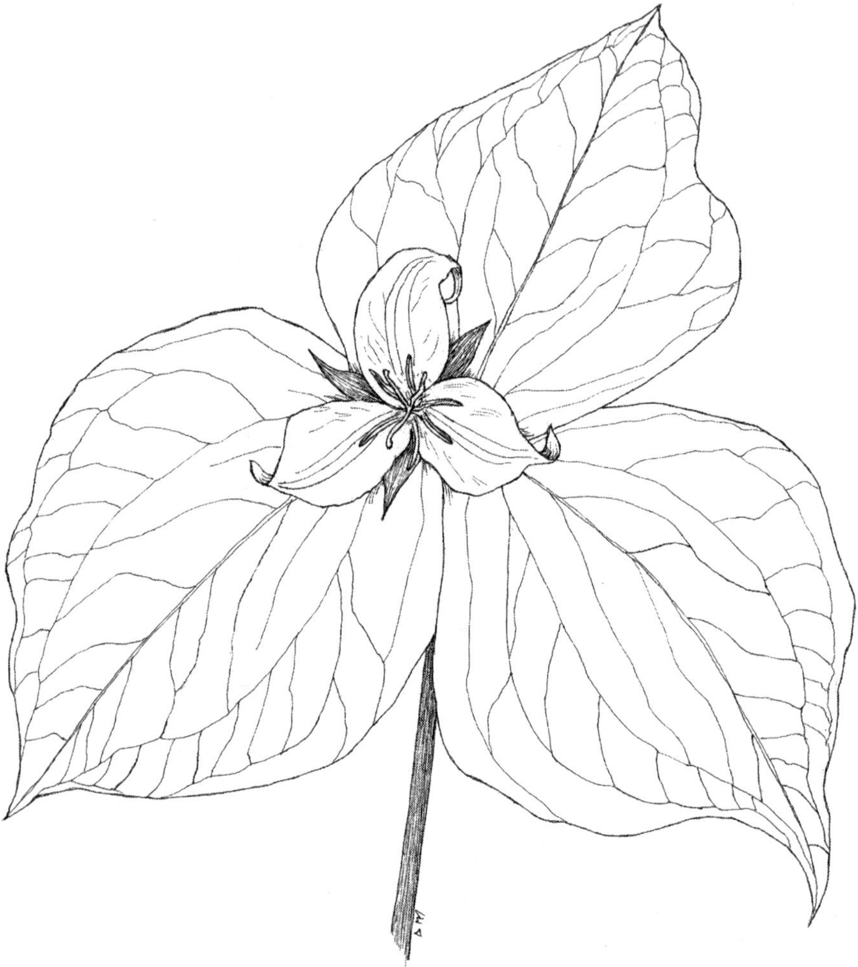

Western Trillium, Western Wake-robin, *Trillium ovatum*

Trillium: the threefold lily,
triple in leaves, in petals and in sepals,
all held on a single stem.
Where the soil is moist in spring
by a seepage or fallen log
it leaps into growth, one day
just breaking the soil, and presto!
three days later in flower
--so fast, it 'wakes the robins'
who shout with startled joy.
The leaves are 'ovate' and ample,
often as broad as long,
coming to a definite point;
they provide a level platform
above which rises the flower,
whose cool whiteness matches
the cool and quiet of the woods.
Wax-paper-thin, its petals
are sometimes almost translucent.
As they age, they turn from white
to shades of mauve, with the flower
still perfect in form, and at last
to a deep and wine-dark purple,
mature and splendid, set off
by the yellow stamens at the centre.

Then its splendour is deposed
by the seedpod full of new life.
Each seed when ripe has a tiny
appendage rich in oil,
attractive to foraging ants,
who lug them back to their nests,

eat the appendage, and discard
the seeds--an effective dispersal.
But none of this can happen
if the flower is picked. Even worse
if the stalk and leaves are picked,
for then the bulb may die.
Let us lay a curse on pickers!
If there were justice in the world
it should be the picker that dies,
not the plant that is picked.

Calypso Orchid, Fairyslipper, *Calypso bulbosa*

In myth Calypso was a nymph immortal
with lovely hair. Her name in Greek
suggests concealment: she lived remote
on a wooded island, her home secluded
by alder and poplar and fragrant cypress.
So the Calypso orchid hides
its loveliness amidst the moss
in the deep shade of the evergreen forest.
At first you would think it a Shooting Star
on a shortened stem, with the same flared petals
and the same rose colour. But the lowest petal
is swollen like a lip, and striped and speckled
with hues of sienna and purple-madder,
exotic and tropical—less like Calypso
and more like Circe the sultry temptress.
Like other exquisite hidden things
Calypso is threatened by the very approach
of hulking humans. For its tiny bulb
is held and sustained by delicate roots.
Any attempt to pick the flower
will snap the roots and starve the plant;
and a careless blow from a hiking-boot
or mountain bike will surely doom it.
The orchid, then, is all too mortal,
unlike the nymph. And yet each year
in shadowed quiet untrampled places
it unfolds in May a timeless beauty
like spring itself, new and eternal.

Salal, *Gaultheria shallon*

In 1906 the *Valencia*, out of San Francisco,
was heading north with passengers, bound for Seattle.
In the fog and blindness of a January storm
she missed the entrance to the straits, and was swept by currents
onto the rocky coast of Vancouver Island,
Graveyard of the Pacific. She lodged on a reef
not a hundred yards from shore, yet far from safety
with no human help at hand, and faced with a line
of sheer unscaleable cliffs. The pounding waves
began to break the ship open, while passengers and crew
huddled on deck, or clambered into the rigging.
Once news was received, rescuers on land were slowed
by the massed vegetation of the deep rainforest,
where storm-felled logs lay crisscrossed over each other
and the wiry stems of salal formed tangled thickets
over head-height. One party struggled to the cliff-top
on the third day after the stranding, but only in time
to witness, abashed and powerless, the final disaster:
a mighty breaker demolished the ship's upper structure,
and pitched the last survivors, some forty souls,
into the surf, where they perished.

And yet for those peoples who had lived and thrived on the margins
of ocean and forest for countless generations,
salal was an aid to survival. Its deep-blue berries,
pressed into cakes, were a staple food, much prized,
and its glossy leaves when young could be chewed by a person
lost in the woods, to relieve the pangs of hunger.
(John Scouler, touring the coast in the 1820s,
wrote of "penetrating [sic] into those primeval forests
never before explored by the curiosity of the botanist,"
with salal "the first object [sic] that attracted our notice"—

as if there had not been intimate, living knowledge
for aeons of the virtues of plants, of the special locations
where particular forms were found, and of how these gifts
of nature might be sustained, while sustaining humans.)

With the wilderness partially tamed, we have the leisure
to enjoy the beauty and vigour of what remains.
Walk the trails in the milder weather of May:
you find strong contrasts of colour in the mounds of salal.
The dark-green older leaves set off pink leaf-buds,
and pale new leaves, and white racemes of flowers
often hanging in parallel rows, like beads or raindrops.
Spotted with pink, they are shaped like inverted urns,
well adapted to shed the drenching rains of their home.

But with access comes the lust for exploitation.
Nowadays even salal is commercialized,
commodified, turned into 'product' - as if produced
by any power other than nature's. Its handsome foliage
is employed as a backdrop for bouquets in the 'floral trade'.

> Spark their imagination and touch their hearts with our memorable
> Valentine's arrangement!
> Carnations, Asiatic lilies, daisy poms, waxflower and salal, designed
> by our select florists
> Stunning pink fluted vase
> A terrific gift idea that sends just the right sentiment this Valentine's,
> at a fantastic value you'll love! Components may vary.

When did we start to speak in this simpering way,
in language that displays its falseness? What better to "spark
their imaginations" than the glory of living plants
not "designed" but in careless abundance in their right surroundings?
And what could "touch their hearts" more truly than words
not bought or sold, but struggling to speak the truth
and to render thanks for all that is freely given?

Dogwood, *Cornus nuttallii*

Beneath the mottled grey bark
the wood of this tree is hard.
It was used in an earlier age
for arrows, bows, harpoons
and for skewers and dagger-shaped tools
called 'dags', whence dagwood, dogwood.
    In winter you would think no life
subsists in the knobbly flowerbuds.
But as the days grow longer
the flower-bracts lengthen out
till they form a ring of 'petals',
overlapping, broad and creamy,
surrounding the small true flowers
clustered together at the centre.
    A young tree set in full sun
will be clothed completely in bloom;
but for my part I prefer
an older forest tree
that has grown into its place
and spaces its tiers of leaves and blossoms
    just so in the woodland openings
        to catch the falling light.

But now that grace is threatened,
as many exquisite forms of life
are threatened. Since the 1980s a plague,
more silent than men with chainsaws but not
less deadly, has ravaged our native Dogwoods.
Caused by a Discula fungus, it appears
on the opened leaves as spots and blotches,
tan in the centre with purple margins.
The infected leaves droop and wither

but cannot fall. Then malformations
begin to be found in the bracts of the flowers,
and lesions on the berries; cancers unfold
on the stems and branches. As spring returns
the fungal spores ooze out in clusters,
and are spread by birds and splashing rain.
The tree may send out sprouts from its trunk
to replace the dying limbs, but the fungus
quickly invades them. Cankers attack
the trunk itself, and its bark buckles.
Within a few years the tree is dead,
or dead in part, its crown withered.
Meanwhile the plague moves on like a flooding
tide, and leaves few trees unscathed.

It seems a paradox that trees
for all their hardness, strength and size,
are so susceptible to attack
by spores of fungus. Mighty elms
whose ancient branches offered shade
are now laid low; and Lodgepole Pines,
tall and straight-grained, have turned in millions
from green to the red of death, destroyed
by pine-bark beetles that carry fungus.
The forces of destruction battle always
against the forces of vital growth,
Strife against Love, and in our age
destruction gains the upper hand.

Twinflower, *Linnaea borealis*

This diminutive plant is a favourite of many
for varied reasons. Some admire those miniature
trumpets, the pale-pink flowers, that swing
three inches from the ground or a carpet of moss,
like congenial twins, one on each end
of the Y-shaped stalk. For others the scent
that enriches the air on still May evenings
is the chief attraction, blending English honeysuckle
with Lily-of-the-Valley. For my part I like
the leaves, so glossy you would swear they were wet,
growing in a series of companionable pairs
on the trailing stems; being evergreen
they show up at all seasons, and their shining surface
catches and gladdens the wanderer's eye.
Others love the plant for its connection
to 'boreal' lands. It grows in the North
from Sweden and Lapland around the Pole
to Russia, China, Scotland, Greenland,
Alaska, Canada, and several of the States.
So many an Ausländer has been cheered to find
this native of the Old World rooted in the New.

Linnaea is named for Carolus Linnaeus,
Prince of Botany and Prince of Naming,
who devised our two-name classification,
(first name for genus, second for species),
and organised the created world:
"Deus creavit, Linnaeus disposuit,"
as he liked to say. Not a modest man,
yet he loved this modest plant for its presence
in Småland, his home in southern Sweden,
and its presence in Lapland, where he saw it as he journeyed

in the ancient forests. *His* name derived
from the Linn or Linden, the warden-tree
of his family's farm: and so it descends
to this friendly plant, not exactly a warden,
but a cheering companion on northern trails.

Bunchberry, *Cornus canadensis*

It grows in moist places, where its roots can ramble
through the spongy substance of fallen logs
to be nourished and watered. Its leaves are handsome,
their "dogwood veins" running parallel
with the edge of the leaf. They expand in a bright-green
whorl or ring near the end of the stem,
setting off the cream-white flower above them,
hand-high from the ground.
A contrast of form as well as colour,
for the leaves make a circle, while the flower's four petals
are sometimes so broad as to fill out a square—
a Maltese Cross with its points reversed.
As in Dogwoods, the Bunchberry's giant cousins,
these so-called 'petals' are really *bracts*,
i.e. modified leaves. The genuine flowers
are tiny, bunched together at the centre,
a miniature thicket of mauve and white.

By Fall they have turned to vermilion berries,
still 'bunched' together, with the leaves as foils,
now bronze and purple.
But 'Bunchberry' is much too clutzy a name
for this well-bred plant. Its cream-white bracts,
their breadth, and the breadth of the deep-veined leaves,
suggest a Confucian serenity.
As the Dogwood flower is the Province's emblem
so this small *Cornus* seems emblematic
—of joie de vivre, perhaps, of contentment,
or simply of being itself,
of being at home in its native place.

Vanilla Leaf, *Achlys triphylla*

Often in the forest you will find a carpet
of rounded yellow-green leaves. Among them
are slender spikes of cream-white flowers
like miniature torches in the woodland gloom.
Each leaf is held a foot above ground
by a wiry stem sent up from the rhizome.
Its shape: a circle with scalloped edges
and deep clefts running like spokes to the centre,
splitting it into three lobes or leaflets.
To children's delight, if you take out the central
lobe, the remaining two by magic
transform to a pair of angel wings.
First and last the leaf is ghostly.
In spring it emerges furled, an almost
colourless membrane, a delicate fan;
in fall, when the leaf itself is gone,
its veins persist as a lacy network.
Each leaf contains an elusive fragrance:
drying, they release a faint scent of vanilla
or new-mown hay. Once country folk
would collect them and hang them in bunches to perfume
the house and ward off flies and mosquitoes,
and they named the plant Sweet After Death.

Tiger Lily, *Lilium columbianum*

Bright orange, mottled with deep-red spots—
but shouldn't a tiger have *stripes*?
It should, but the name 'Leopard Lily'
was already taken by a garden species,
so 'Tiger' it had to be.
Yet the connotations are right,
for it adds a touch of tropical splendour
to the temperate woods of the coast.
        Six long petals (properly *tepals*)
compose the flower.
When the time is ripe, they begin to open
from the tip, to form an attractive bell
—but then go on curving, curving,
as if pulled back by a tightening muscle,
till they meet themselves at the base.
On a hot June day I have watched a bud
opening at suppertime, a bell by mid-evening,
fully curved back by dusk.
And such exotic forms!
Each recurved tepal resembles
the toe of a caliph's slipper,
while the six together suggest a crown
of reddish gold and Eastern richness.
        Now is your chance to examine
the inside face of the flower:
glossy bright and brilliant orange,
lightening to a luminous yellow-green
at the base, around the nectary.
You will see the spots are maroon,
not brown as appears at first,
and often they run in lines,
as if sprayed from a loaded pen.

By now you may be distinguished
by gold-dust brushed on your nose,
for the anthers stand proud of the flower
as much as two inches. When fresh
they are fat and red, but soon
shrunken and chocolate-brown,
after yielding their plentiful pollen.
A typical plant will have four or five flowers
on an upright stem perhaps waist-high.
But on the banks of Gordon River
in deep moist soil
I saw a plant with twenty blooms,
way over head-height,
towering among the rampant salal.
You will find this lily again
in the cool damp soil of alpine meadows
with a single flower, dwarfed by the climate
—always encountered there with surprise
as a well-known face from another world.

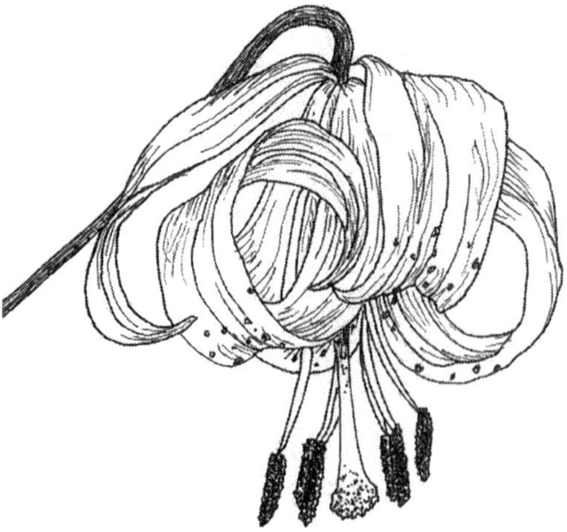

Devil's Club, *Oplopanax horridus*

The Devil has several
evil plants:
the Devil's Horn
is the stinkhorn fungus,
Phallus impudicus;
his Milk, the poisonous
Sun-spurge juice;
and asafoetida
with its stinky gum
is the Devil's Dung.
Old Nick's 'weapon'
(*oplon/hoplon*)
lurks amid black
and mucky soils
in cedar swamps.
Its crooked twisted
man-high stems,
sheathed in gray,
are spiked with prickles.
Stiffly they reach
to seek their prey;
if you trip on a root
cunningly placed
close to the ground,
they will catch at your face
or pierce your clothes,
and the wounds will sting,
red and inflamed.

In winter the stems
look dead, or waiting,
but in spring the lurking
life emerges in
monstrous leaves,
as the tips of the phallic
stems extend
and explode into sprays
of cream-coloured flowers.
First-peoples felt
the potency of the plant,
and painfully forced it
to serve their needs.
With its snagging power
they hooked the octopus
from his rocky lair;
they blackened faces
with the charcoal powder
to ward off evil
in the ritual dance;
made poultices
of the inner bark
to counter pain;
and steeped the stems
with the thorns removed
for a medicine sovereign
against many ills.
And so they supped
and diced with the Devil,
and tweaked his nose.

Fireweed, *Chamerion* (or *Epilobium*) *angustifolium*

Wherever the woods are broken, the earth laid bare
by a forest fire that lights the trees like torches
and leaves them stark and blackened, or an avalanche
that tears a glistening rent in the mountainside,
or a fiery volcanic blast on Mount St Helens,
but chiefly where humans have scarred the hills with roads,
clearcut great swaths, and burnt the piles of slash—
Fireweed quickly invades such clearings as these,
setting them alight with the blaze of its rose-red flowers,
with healing fire this time, as it clothes the soil.
A plant of water as well as fire: it likes wet ground,
and in England they call it Rosebay Willowherb
for its long thin leaves like those of the Weeping Willow,
a moisture-lover.
                                        As you walk through the crowding spires
of flowers, observe how they ripen in ascending order:
seedpods already formed below, then flowers
at the centre of the spire, and unopened buds above.
The seedpods, long and slender, shaped like the leaves,
are crammed with seeds, each tufted with silky hairs—
so abundant that the silk was mixed with scarcer wool
by the Saanich and Squamish people, and used in spinning.
These tufted seeds can be carried for miles on the wind
like an airborne invasion, a miniature paratroop army
dropping out of the sky; next year they spring up in hosts,
and the damaged earth is suddenly alive once more
with greens and reds and the sounds of foraging bees.
                              In London's Dockland, where I lived as wartime child
in the 1940s, firebombs rained from the sky
at the time of the Blitzkrieg terror, and V-1 rockets
delivered their deadly payloads. Everywhere
there were blackened sites where houses had just now stood,

where children had played and pensioners drank their cuppas,
where housewives hung out the washing, and their working men
at the weekends tended their lettuces, sprouts and onions.
On these sites too the fireweed blazed, transfiguring
what people had laboured to build and to destroy.
I wonder if it will flower after Armageddon,
clothing the scorched, irradiated earth
unseen by humans, after the final fire.

Harebell, *Campanula rotundifolia*

Part 4:
MOUNTAINS

Mountain Arnica, *Arnica latifolia*

This genus rejoices in a single name,
Arnica in English, Arnica in Latin,
with no confusing common name
(or multiple names) to muddle the issue.
And the Arnicas single themselves out clearly
among the many 'damn yellow flowers'
of daisy form in the Composites
by their leaves that grow in opposite pairs,
not alternating, up the stem—
a sure recognition guide, that soon
leaps to the eye on mountain walks.
The foremost Arnica in coastal mountains
is *latifolia*, the 'broad-leafed' species.
Its flowers are a strong, straightforward yellow,
with petals often blunt and toothed,
set off by a light-brown disk at the centre.
They splash the subalpine meadows with colour,
and brighten the shadows of the scattered trees.
Standing alone, the plants seem ragged;
they look their best amid a profusion
of other plants growing in riot around them,
pink heather, perhaps, and blue-white lupines,
and bunchberry carpeting the ground at their feet.
On a high trail in Manning Park
on a day of crisp and brilliant sunshine
I met a grizzled man of the mountains,
lithe and weathered, of age like mine.
"Take the flowers of Arnica," he said,
"and steep them in alcohol, say vodka,
for a couple of weeks—a powerful tincture
to rub on the skin against inflammation,
bites and bruises and weary muscles."

Then with a rueful glance he added,
"Thank God for the sun and breeze and these meadows, and the strength to enjoy them: we shall not be walking these trails for many more years, you and I."

Harebell, *Campanula rotundifolia*

These well-formed bells, diminutive
(campan*ula*) but full and rounded,
swing in the breeze on slender stems
like sounding bells in a campanile
somewhere on a hillside in Tuscany.
Church bells call us to weddings or prayer,
but these are the bells of memory:
tokens of the Old World found in the New,
they recall for many the scenes of childhood.
"Truly the sight of these fair flowers
in lonely spots in Canada,"
wrote Catharine Traill, 1869,
"has carried one back in thought
to the wild heathery moors
or sylvan lanes of the mother country."
Growing throughout the northern world
Harebells have adapted to differing sites,
differing climates, differing soils;
every country, almost every alp
produces a subtle variation
in breadth of bell or angle of lobe
or density of the silver down.
Still more diverse are the colours of the bells:
in Scotland a pale blue form prevails,
elsewhere a deeper blue, or violet,
recalling the sheen of Satinflowers;
pinks and purples are also found,
pure whites, and delicate silver-pales.
Perhaps it is the plants' resilience
that endears them so to mountain-lovers;
those slender stems, thin as a hair
(whence Hairbell, Harebell) look too frail

to survive the storms, yet thrive by yielding—
bow to the gale, but do not break.
    Most people know the *summer* leaves,
long and skinny, but Linnaeus named it
the 'round-leaf' plant, because he passed it
daily beside the University steps
in *winter*, when the leaves are round or oval—
an emblem of Polanyi's truth, that knowledge
is always personal, however precise,
and shaped by experience, memory, and love.

Lupine, *Lupinus arcticus* (also called *latifolius*) and other species

Lupines with their blue-white spires of flower
are so familiar in suburban gardens
that they cause surprise, a clash of context,
when seen in the untamed mountain world.
There they provide the dominant blue
in the flowering meadows, and among the trees;
sometimes you find whole lupine fields,
sweet-scented, rivalling the sky for colour.

Why is it that flowers atop the spire
look whiter than those towards the base
that opened sooner? Recall the structure
of pea-type flowers: a *banner* petal
flaunting above, and two side *wings*
enclosing the *keel*. In lupines the banner
is white-and-blue, and at first is tightly
folded over the rest of the flower
as a cover from rain, so the white shows up;
later the solid blue of the wings
predominates, after the flower has opened.

The leaves of lupine have a special beauty.
They are *palmate*, viz. cut into leaflets
that radiate out from a central point,
like the leaves of a palm, or Queen-cup petals,
or the rays of a star. Their deep-cut shapes
make intricate patterns of light and shade,
constantly changing as they track the sun.
The leaflets are grooved, to collect the dew
and channel it into the heart of the leaf.

Beneath these droplets sufficient air
is held by the tiny hairs on the leaf
to make the water sparkle like a jewel;
and so at dawn each leaf is transformed
to an ornament, diamond-set at the centre,
such as no craftsman could hope to equal.

Fan-leaf Potentilla, *Potentilla flabellifolia*

This is a 'potent' genus,
potent in medicine for aeons
in Europe, and potent in colour.
    In our coastal mountains, the most
abundant of the Potentillas
is the Fan-leaf species, named
for its leaflets (they come in threes):
each is a little fan-like
(if you are fanciful),
with two straight sides diverging
from the stalk in a V, and the outer
edge in a semi-circle,
deep-toothed—a raggedy fan!
    Often this plant creates
an understorey of yellow
in the meadows, with a carpet of phlox
beneath it, lupines above,
and valerian forming the roof.
If Arnica too is present,
they are easily told apart:
Potentilla has just five petals
forming a cup-shaped flower,
whereas Arnica's flower is a star,
with at least eight petals for rays.
In colour, Potentilla's
gold-yellow shade is deeper
than Arnica's straight yellow.
    More care is needed to distinguish
Potentilla from Buttercup,
since both have five-petalled flowers,
so you need to check those petals:
Potentilla's are notched

quite deeply on the outer edge,
not so with Buttercup.
Still not entirely certain?
Turn the blossom over.
If on the bottom there are five
little sepals, yellow, be sure
you're holding a Buttercup;
if there are ten green sepals,
this is a Potentilla.
In the crystal air of the mountains
they contend in richness of colour.
Buttercup's flowers are known
for their shining yellow: the cells
on the petals' surface are smooth,
and reflect the light, creating
a waxed or varnished appearance.
But in Potentilla the cells
present a roughened surface,
which yields a deep matt colour,
a velvety golden-yellow,
a potent shade, beside which
all other yellows pale.

Silky Phacelia, *Phacelia sericea*

Coming down the trail from Klahhane Ridge
I rounded a bend and stopped, surprised
at the sight of a plant not seen before
in a summer's hiking. It was bold and handsome,
with flower-spikes seven inches long,
lavendar-purple, and silvery leaves
smooth to the touch, with scalloped edges
like Dusty Miller. Startled as I was,
I took some seconds to connect these clues:
Silky Phacelia (sensuous name)!
I bent to admire the special feature
of this plant, its long protruding stamens
almost like hairs: their golden anthers
set off the royal colour of the flowers,
an effect like Camas. But why was it found
just here? The place itself gave answer:
it was growing beneath the sheltering branches
of a Yellow Cedar. Going on my way
I began to notice other plants
not seen in the open meadows, but gathered
under the shelter of firs or cedars:
Black Gooseberry and Columbine,
Evergreen Huckleberry, Heuchera,
Coast Penstemon, even Bleeding Heart,
all enjoying the trees' protection
against the wind and sun and snow.
So one observation leads to another,
and if we are mindful enough and alert
the world around us opens itself
to our eyes in sudden revelations.

Trollius laxus

Marsh-marigold (*Caltha* species) and Globeflower, *Trollius laxus*

These are plants that relish wetness,
squelchy meadows, the banks of streams.
They are cousins, members of the Buttercups,
Ranúnculáceáe (which means
'the little frogs', i.e. moisture-lovers).
Both are precocious, poking their way
through melting snow and icy water,
exuberantly ushering in
the flowering season, then setting seed
while other plants are still a-bloom.
Their flowers are similar, cups of white
with a showy centre of yellow stamens;
Caltha typically has eight petals,
Trollius five. But the leaves are unlike,
though handsome in both: in Caltha *entire*
(i.e. not divided) and waxy-shiny;
in Trollius soft and much-divided,
forming a ruff around the stem
below the flower, to set it off.

Now to naming, that powerful magic.
Caltha may come from *kalathos*,
calathus, meaning a bowl-shaped basket
or a cup, a goblet. 'Globeflower' tells
that the flower takes on a spherical shape
when it closes up in frost or rain.
Trollius is just a Latinisation
of an old Germanic name, *Trollblume*,
but no-one knows the meaning of *that*.

Let's indulge the notion that it means 'Troll-flower,'
and derives from the fact that Trollius grows
in squishy places under bridges,
which is where the fierce Troll lurked, before
he was trounced by the biggest Billy Goat Gruff.

Spreading Phlox, *Phlox diffusa*

Phlox: the word means 'flame.'
This is a plant that thrives
in open places, and most
in stone and broken rock,
where it can gaze directly
at the bright flame of the sun.
Down through the rocks, halfway
to the ocean it sends a taproot
to mine some tiny vein
of moisture against drought
from the fiery sun and the wind.
Above ground, stubby leaves
in a mat just two inches high
store that life-giving moisture.
The sunniest weeks of the year
cover these mats with flowers
held close against their surface.
How to speak of them justly?
Some praise their pastel colours
ranging from white to palest
pink to lavender blue
and on into mauves and purples—
all set off by the golden
anthers poised at the centre
as if in a tiny cup.
But a more distinctive aspect
than colour is the form of the flowers,
crisp, clean-edged and perfect.
Specially classic are those forms
where the petals are shaped like a wedge
with two straight sides, not touching
or overlapping: then the channels

running between them show dark
from the shaded leaves below.
Here light and dark interact:
the dark takes on its own form
defined by the petals' flame.

Penstemon, especially Davidson's, *Penstemon davidsonii*

Since *pente*, 'five', begins the word Penstémon,
pentameter is the only choice for this poem.
'Five stamens' are inscribed in the name of the genus,
and five it has (an excessive number for a member
of the Figwort family), but only four bear pollen.
Penstemon flowers have a trumpet shape, ideally
evolved over time for pollination by bees.
Deep folds run back along the trumpet's stem;
the pistil presses into the topmost fold,
while those five stamens arch against the roof.
The mouth of the trumpet opens to five bright lobes,
two tilting up, while the other three slope outwards
to create a level strip. The bee lands here,
and crawls inside the trumpet, where her furry back
brushes against the overarching pistil,
quickening it with pollen from other plants,
while the stamens drop more pollen on her back, to be carried
to neighbouring flowers. Her reward for this performance
is the nectar waiting at the end of the trumpet.

There are many varied species of Penstemon,
some tall, two feet or more, and others low;
some evergreen, some not; some with small flowers
a centimeter long, some three times longer.
And several species range into the mountains,
especially the sheltered valleys. But one of the truest
mountain-dwellers, at home on alpine ledges
and on arid slopes of broken rock and gravel
is the plant we label Davidson's Penstemon
(absurdly, since it evolved its perfect form
aeons before the Davidsons were thought of).

Against the ferocious winds of its highland home
it forms dense mats, just inches above the ground,
and its leaves are short and leathery, with thick edges,
to conserve the precious moisture. But its flowers are lavish
in the length of their trumpet stems, in abundant numbers,
and in the wine-rich quality of their colours,
a splendid mauve, or lavender-blue, or purple.
Silky-textured, they seem to be luminous
with the special intensity of mountain light.
Barely ten weeks of summer intervene
from snow to snow on the upland slopes and ridges,
yet this plant thrives in its solitary fastness,
as if self-delighting, enjoying its own perfection,
though it may remain unseen and unadmired
by any human eyes from year to year.

Paintbrush, *Castilleja* (various species)

Plant as paintbrush: metaphor here
at work once more. Legend tells
of a Blackfoot woman who in mortal danger
used her own blood to paint a picture;
where she dropped her twig, up sprang this plant,
and hence its usual colour red
(though it can be cream or orange or purple).
The showy petals that compose the brush
are really coloured leaves, or *bracts*,
like the sturdy 'petals' of a Dogwood blossom.
Poking out from the bracts, or hidden among them,
you will find a long thin beak or probe,
yellowish-green, often edged with red.
This is the flower, believe it or not—
more precisely the upper lip of the flower
(the botanists use a different metaphor,
the *galea* or helmet). If you open the slit
on its underside, there tightly packed
are the stamens and pistil.
No problem at all
for hummingbirds to slide their beaks
down through that slit, and into the nectar.
Like other red plants such as Columbine,
Paintbrush has evolved for pollination
by hummingbirds, because the birds
are able to perceive the colour red,
while bees cannot. In fact the birds
and plants have probably *co-evolved*,
the 'beaks' or spurs of the plants growing longer
as the beaks of the birds grew longer too.

When the tiny seeds of the Paintbrush flower
are cast on the ground, and seedlings sprout,
they must quickly send down roots, and link
with the roots of nearby herbs and grasses,
or else they perish. Once they have grown
green leaves, they can photosynthesize,
but they still draw nutrients from their hosts.
For this they are labelled parasites,
even scolded as moral reprobates,
but in fact there are such dependencies,
such links, throughout the natural world,
plant to bird, plant to plant,
and in mycorrhizal associations
of plant to fungus. As John Muir said,
"When we try to pick out anything by itself,
we find it hitched to everything in the universe."

Pink Monkey Flower, *Mimulus lewisii*

A mass of rose-pink flowers,
more vivid because they are seen
by the brightness of ice-cold water.
The stems, knee-high or close,
bear handsome pointed leaves,
strong-veined, but soft to the touch,
covered with tiny hairs.
But the flower is the thing: a long
and showy trumpet, opening
into five brilliant lobes.
Pressed in the roof of the trumpet
are two small plates like electrodes,
the flattened ends of the pistil,
placed so that any insect
is bound to brush against them
as it enters the trumpet-stem.
(As in all the Mimulus tribe,
these plates are sensitive, and will close
once touched by an insect or bee,
shutting in the life-giving pollen.)
Two ridges on the floor of the trumpet
are handsomely marked with deep yellow,
and with lines of red dots, as bee-guides.
Further in, the trumpet turns white,
and last, in the nectary,
a luminous lemon-yellow.

As summer matures, the flowers
begin to drop, still perfect,
into the streams and ponds
alongside which they grew;
they form bright rafts of colour

drifting down the creeks,
and turn small pockets of water
into natural fingerbowls
whose beauty far surpasses
the rose-petalled bowls of kings.

Queen-cup, *Clintonia uniflora*

"This is a boring plant," says Jack.
"Its leaves are dull, too uniform,
just plain old green. And what kind of name
is 'Queen's Cup'? Any fool can see
that the flower is shaped like a star, not a cup."
"Wise guy!" says Bill. "Your problem is
that you want excitement and vivid colour
from every plant. But Queen-cup has
a modest beauty all its own.
Those "plain green leaves," if you look up close,
are soft and glistening like silk.
As for the name, don't blame the flower!
Anyway, *Queen-cup* is the proper form,
not Queen's Cup: Julia Henshaw named it
as the queen of all the snow-white flowers
in the mountain woods."
                                        "These royalists!"
says Jack. "The 'queen' is an ice-queen, then:
she can only manage a single flower!
Those upright stamens are much too tidy,
the petals are too straight-edged, and there's something
frigid about that dead-white tone."
"Your problem is that you have no soul,"
says Bill. "She's more like a spirit of the place:
the fresh cool quality of this lily
matches the chill of the mountain air.
The flower has a quiet perfection, single
and self-delighting; no need here
for elaborate form or extravagant colour.
Where whiteness reigns for much of the year,
white befits this woodland queen."
"Want to know what *your* problem is?"
says Jack …

White Avalanche Lily, *Erythronium montanum*

Yellow Glacier Lily, *Erythronium grandiflorum*, and White Avalanche Lily, *E. montanum*

These are twin sisters, mountain-dwellers.
They are much alike, and differ chiefly
in their coloration. The yellow lily
has a golden hue when seen up close,
but takes on more of a lemon tone
when viewed en masse as it cascades
down a slope, an avalanche of colour.
The white flower shows its kinship to its sister
in its golden centre, sharply edged—
a startling birthmark, a family emblem.

The grace of these lilies belies their strength.
I think of them as an Amazon army,
the shock troops of advancing summer.
They spring from the icy ground in hosts
like the warriors sown from the dragon's teeth,
and follow up the retreating snow
implacably, with the sun their ally.
They flourish in nature's turbulence,
the heaving of ground by frost and ice,
rough digging by bears delving for food,
and they thrive especially when the soil is disturbed
and cleared by fire or avalanche,
for this lets sunlight reach the soil
and releases nutrients to feed their bulbs.

First nations knew it, and would set cool fires,
under control, to clear the brush
and encourage the growth of the yellow lilies
as well as other bulbs and berries
on which they relied for staple foods.

With the white man's coming, the fires were banned,
the brush grew dense, the lilies dwindled,
and a way of life attuned to nature,
and following nature's methods, was ended.
Nature is not static but ever-changing,
sometimes by violence. If we would live
sustainably, under nature's guidance,
fire is one of the lessons taught.

Western Anemone, *Anemone occidentalis*

With snow still on the trail
and the yellow lilies blooming
as summer comes to the mountains,
you will find white cup-shaped flowers
just inches above the ground.
They challenge you as the stranger:
"Before you pass, who am I?"
Caltha and Globeflower are options
already considered above;
but if the stems are hairy
and the leaves finely divided,
then answer boldly "Anemone!"
and pass unhindered on your way.

Anemone is among the first
to flower in the high country
and the first to fade, yet its seedheads
last on throughout the summer.
The flowers open in haste
as soon as the plant emerges,
while the leaves are still tight-furled
and pale, as if shocked by the light.
In these first days the stems,
even as they carry the flowers,
continue to grow, and the leaves
unfurl, most at the base,
but some in a ruff round the stem
just under the flower. Too soon
the petals (really sepals)
begin to brown at the edges,
crumple and fall. But now
there appear those shaggy heads

of plumes, each fused to a seed;
the effect is an upside-down
dishmop, or a human head
with a shock of tousled hair.
Some call them "Towhead Babies"
from the silky softness of the plumes.
Later the hairs become worn
by sun and wind and rain,
showing their age, and the heads
grow bald at the crown. Even so,
they make a brave display
in the sere late-summer meadows—
especially when lit from behind
by the slanting rays of the sun—
until they are beaten down
by the first snows of winter,
and all is whiteness once more.

A Riddle

Riddle me, riddle me, riddle me ree.
I am an enigma, a paradox,
a mountain mystery: here are some clues.
I bear deciduous leaves, and yet
I belong to a genus with evergreen leaves,
thick and leathery. I enjoy the shade
of coniferous woods, yet I thrive as well
in sunny locations, out in the open.
My flowers are handsome, well-formed cups
an inch across, white or creamy,
in showy clusters, a delight to the eye;
yet hikers call me 'mountain misery'
because I spread in tangled thickets,
blocking their path. My Fall display
of bronze and orange and red would grace
the Eastern Townships, yet here I am
among these ever-green firs and cedars.
Give up? You say I'm a rhododendron?
That's not fair—you peeked at the answer!

The full answer is in the endnotes.

Broad-leaved Willowherb, Broad-leaved Fireweed, River Beauty, *Chamerion* (or *Epilobium*) *latifolium*

This is the loveliest of our mountain flowers
(give way, Penstemon and Phlox and others);
even people immune to the beauty
of other flowers are awed by this.
If you miss it here, make a pilgrimage
to the Yukon to see it in all its glory.

As yet, like other distant beauties,
it eludes our naming. 'Willowherb' suits
its taller cousin, the Common Fireweed
or Rosebay Willowherb, whose skinny leaves
are willow-like—but not this plant,
whose leaves are broader; yet hardly 'broad',
being only a third as wide as long!
And 'Fireweed' is a name that fits the cousin,
which springs from the blackened sites of fires,
but not this plant, which grows by the banks
of icy streams. 'River Beauty' works
in the North, but here in the South it appears
by mountain creeks, not lowland rivers.
Surely 'Rose' should surely be part of the name
to do it justice—but what's in a name?

Though the plant is barely a foot in height,
its flowers are bigger than its cousin's, and seen
more clearly, because less massed together.
They display a geometric beauty,
a perfection of form. There are four broad petals
that alternate with four thin sepals:
a broad square cross, as in Bunchberry,
with a narrow cross bisecting its angles;
all this enhanced by colour, pale purple
with deeper purple veins in the petals.

The flowers, as the mountain sunlight streams
across them, glow with a luminous brilliance
at one with the dazzling brightness
of mountain gravel and mountain water.

Sitka Valerian, *Valeriana sitchensis*

Valerian is told apart
from several look-alike plants,
that have umbels of small white flowers
atop their stems, by two
diagnostics: stamens and stink.
Its stamens stand quite proud
of the flowers, giving the umbels
a fuzzy appearance, a halo.
Yarrow is common, but it lacks
those telltale stamens, and its leaves
are fernlike; Valerian's aren't.
The scent of Valerian flowers
has varied effects: so powerful,
"it could give a hiker an attack
of giddy ecstasy,"
according to Harvey Manning;
others just think of washrooms
and pee. But the root, at least,
is universally agreed
to reek of unwashed socks,
strangely attractive to cats.
One Valerian species
was known to the Greeks by a most
unGreeklike moniker, *Phu*,
surely an expression of disgust.
Tincture of Valerian root
is used as a sedative
and an aid to sleep: as in Buckley's,
medicines are best when nasty.

The source of the name is unknown,
but it has a consonance
with Latin *valere*, 'be strong,
be well'; the 'wellness' of those
who are helped by the potent root
is matched by the 'strength' of its pong.

Pink Mountain Heather, *Phyllodoce empetriformis*, and White Mountain Heather, *Cassiope mertensiana*

Phyllodoce: four syllables,
with the *e* pronounced as in bee, not silent.
Who was she? One of the Nereïds,
all sea-nymphs, all renowned for beauty.
Cassiope (again the *e*
is voiced, not silent) was a mortal woman
who rashly compared her beauty with theirs,
the immortal sea-nymphs': no surprise
that a sea-monster rose to avenge her boast.
Why these exotic sea-linked names
were imposed on our humble mountain heathers
nobody knows. Truth to tell,
the great affection that these plants inspire
is not for ravishing beauty, but rather
for homeliness, for their reassuring
familiar presence in earth's wild places.
Such beauty as they have can be seen up close,
in the form of the tiny bell-shaped flowers.
In Cassiope, the pure white bell
contrasts deliciously with the red of the sepals
that clasp it like a setting that holds a jewel.
But in Phyllodoce the rose-pink bell
harmonizes with the sepals' red.
And in this plant the pistil extends
beyond the rim—the clapper of the bell!
As you study these details, bees will be buzzing
around the heathers, and popping their heads
in the tiny bells, a perfect fit;
they make an unerring selection of those
with nectar, and bypass others in an instant.

If in high summer you see from the trail
a distant hillside glowing pink,
you will know that the heather is in full bloom,
that the bees are buzzing there in the thousands,
and the bells are swaying there in the millions.

Coda: The Mountains

Ascent is easy, though arduous.
You are high on the freshness of upland air,
eager to enter that other world
of space and freedom, of release from cares.
The scent of pines in the morning;
flashes of white on darting juncos;
distinctive voices of chickadee, raven,
nuthatch, flicker, squirrel. And then
as you reach a vantage-point or treeline,
the sudden unfolding
of range upon range of snow-bright mountains
beyond the flowering meadows at your feet.
Here you can pause, drink water, breathe,
and wander freely on the roof of the world.
This is a place as close, it seems,
to the sky as it is to the earth beneath.
Snow comes in any month, and clouds
descend in a moment and hide the trail;
but after showers the brilliant light
gives new intensity of vision.
In such a place thought comes unbidden
to the senses first, and then finds words
from the mind's store, perhaps like these:
that the power that paints the meadow, and fashions
the lovely form of a flower, is also
the power that moves the sun and the other stars.

As the day wanes
the skyladder leads you down,
down through the warming levels of air,
down beside the descending streams,
down through the hazy light of evening.

You return to the well-known place
in the everyday world, the place where you live
and have your being, but you carry with you,
rooted in body and mind, remembrance
of another reality, another world.

Glossary

Some botanical terms are clarified by the illustrations which follow. Others are explained in the poems where they occur, or in the endnotes, as indexed below.

Anther	The pollen-bearing part of the *stamen*, see Fig. 1
Banner	See 'Scotch Broom', 'Lupine' and Fig. 2
Bract	See 'Bunchberry,' 'Dogwood,' 'Paintbrush'
Calyx	'Cup' of *sepals* holding the flower
Cleistogamous	See 'Yellow Wood Violet'
Compound Umbel	See 'Spring Gold'
Entire	See 'Marsh-marigold'
Floret	See 'Oxeye Daisy' and Fig. 4
Galea	See 'Paintbrush'
Keel:	See 'Scotch Broom' and Fig. 2
Lip	See Fig. 3
Lobe	Roundish projection of petal or leaf, see Figs. 1 & 3
Palmate	See 'Lupine'
Pistil	The flower's female organ, see Fig. 1
Receptacle	See 'Thimbleberry' endnote
Scape	A flower-stem rising from the ground
Sepal	A modified leaf, part of the *calyx*; see Fig. 1
Spathe	See 'Skunk Cabbage'
Spur	See 'Red Columbine', 'Menzies' Larkspur'
Stamen	See Fig. 1
Staminode	See 'Harvest Lily' endnote
Standard	Another term for *banner*
Tepal	See 'Tiger Lily' endnote
Thyrsus	See 'Red Elderberry'
Umbel	See 'Spring Gold'
Wings	See 'Scotch Broom', 'Lupine' and Fig. 2

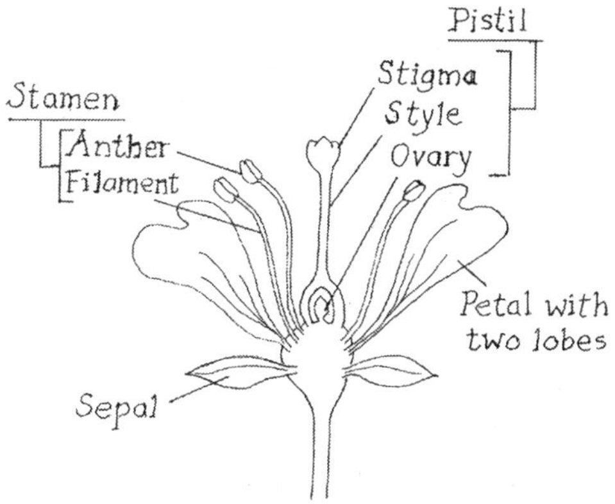

Fig. 1. Cross-section of a regular flower

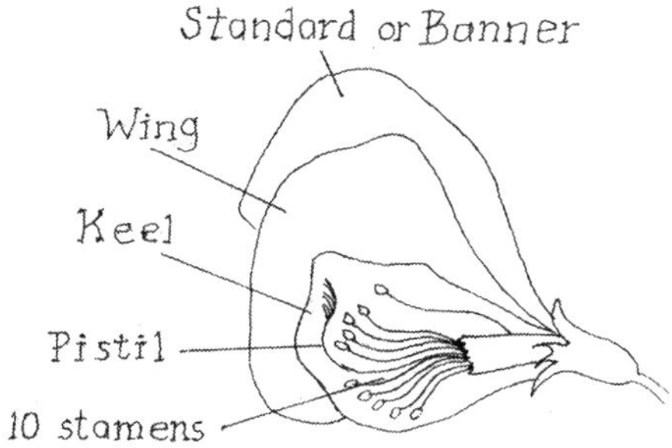

Fig. 2. Cross-section of a pea-type flower

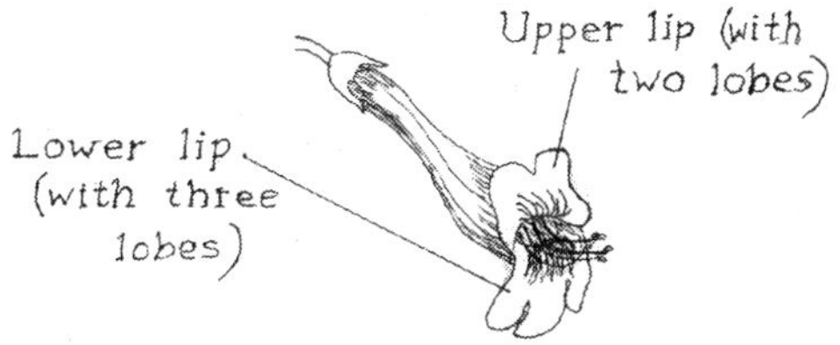

Fig. 3. Tubular flower

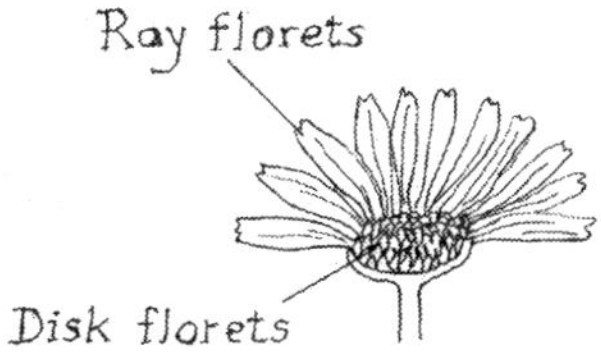

Fig. 4. Composite flower

Endnotes

Overture

The indigenous concept of wildflowers as a blanket for mother earth, and the rich meanings of the concept, are explored in Turner, *Earth's Blanket*.

Part 1: Sea Coast

Rusty-haired Saxifrage

Saxifraga rufidula was formerly regarded as a variety of Western Saxifrage, and named *Saxifraga occidentalis var. rufidula.* Teeth resembling the cogs on a gear wheel: Pojar & MacKinnon 163; Clark 1976 216 says "as if from a cookie-cutter."

Monkey Flower

The likeness to monkey faces is amusingly conveyed in the painting at Ward-Harris p. 69, as well as in Bonnie Moro's drawing. "Gamboge yellow“, "glowing with the sun's own brilliance“: Clark 1976 462.

Mist Maidens

Count Romanzoff was a Russian patron of botanists. I apologize to him for playing on the similarity between his name and that of the Romanoffs.

Spring Gold

Compound umbel: see Pojar & MacKinnon 222 & 507 (illustration).

Arbutus

E.J. Hughes: see Ian M. Thom, *E.J. Hughes* (Vancouver 2002)—or better, when possible, the paintings themselves. For the tenacious roots as typical of Heathers see Penn 161. The Saanich story of the Great Flood is recorded by Pojar & MacKinnon 49, and by Turner *Earth's Blanket* 50-1. The translation of ŁÁUWELNEW as "Place of Refuge" appears in Elliott 65.

Golden Paintbrush

Sunlight lovelier at its setting: Seneca *Trojan Women* 1139-42.

Beach Pea

'Beanpod': the Greek is *phaselos*, whence *phaseolus* is used as the name of a genus in the bean family. 'Raven's canoe': Nancy J. Turner, *Plants of Haida Gwaii* (Winlaw B.C. 2004) 149-50.

Nodding Onion

Cooking-pits: Turner *Earth's Blanket* 198-99, Pojar & MacKinnon 107.

Gumweed

The 'petals' are properly *ray-flowers*: see 'Oxeye Daisy' and Fig. 4. The medical properties of the gum are usually claimed for other species of *Grindelia*, viz. *squarrosa* and *camporum*. However, Haskin 363 claims it for *integrifolia*.

Labrador Tea

The red-brown fuzzy underside of the leaf is well shown at Ward-Harris 35. Haida: Turner, *Food Plants*, 80.

Ocean Spray

Tools and weapons: Turner, *Plant Technology* 182. Nailless barn: Clark 1976 245.

Ancient homes of birds: Vergil, *Georgics* 2.208-09. Felling of the grove: Ennius, *Annals* 175-79; Vergil, *Aeneid* 11.135-38; Statius, *Thebaid* 6.93-106.

The air denuded: Seneca, *Phaedra* 473-74.

Part 2: Field and Slope

Shooting Star

Pliny, *Natural History* 25.28: *dodecatheon vocant, omnium deorum maiestate commendantes.* In his time the name probably referred to the primrose, a relative of the shooting star; since shooting stars are North American, Pliny sadly never saw one. The bee's position in collecting pollen is illustrated and described at Lyons 266.

Camas

Meriwether Lewis: entry of 12 June 1806 in the journals of the Lewis and Clark Expedition, quoted by Clark 1976 14. Tending of fields, harvesting, cooking: Turner, *Food Plants* 42-43. Gathering gulls' eggs: Elliott 23. Nez Percé: Clark 1976 12; Francis Haines, *The Nez Percés* (Norman, Oklahoma 1955) 178-79.

Red Elderberry

Several details and phrases are owed to Clark 1976 496-97. "The liquid fire … ": the famous interpretation (here modified for metre) of the god by E.R. Dodds in his edition of Euripides' *Bacchae* (Oxford 21960) xii. North American uses of Elder stems: Haskin 341, Craighead 184, Turner *Plant Technology* 163-64.

Oregon Grape

Aquifolium, 'prickly-leaf,' has nothing to do with water! The species name was probably borrowed from holly, *Ilex aquifolium*, because of the similarity of the leaf, although the holly family (Aquifoliaceae) does not include mahonia.

Jelly: Pojar & MacKinnon 95.

Strawberry

Bella Coola children: Turner, *Food Plants* 113. Ovid's Golden Age: *Metamorphoses* 1.89-112.

Thimbleberry

The *receptacle* is the end of the stem, to which the flower and later the fruit is attached. Container for berries: Turner, *Plant Technology* 190. Nuu-chah-nulth canoes: Turner, *Food Plants* 124-25.

Pacific Crabapple

Gitga'at: Nancy J. Turner and J.C. Thompson (eds.), *Plants of the Gitga'at People*, Victoria 2006. Nootka: Gilbert Malcolm Sproat, *The Nootka* (ed. C. Lillard), Victoria 1987 (originally published 1868) 43; Turner, *Food Plants* 118.

Oxeye Daisy

Children tumble out of school: Clark 1976 535. Daisy from 'day's eye': *Oxford Dictionary of English Etymology*; Chaucer, *Legend of Good Women*, Prologue 183-5, "That wel by reson men hit calle may / The 'dayesye' or elles the 'ye of day,' / The emperice and flour of floures alle."

Chocolate Lily

Decadent afternoon: Penn 42 ("Temptation in the Oak Meadows"). Lewis Clark's photo: Clark 1976 47. Seed-boxes illustrated Ward-Harris 226.

Mock Orange

Soapy lather: Turner, *Plant Technology* 174. Sam Mitchell: Turner, ibid.

Red Columbine

The resemblance to a covey of doves can be seen in the photographs of Krukeberg 165 and Davies 132. The Japanese word *ma* refers to space, particularly the space between parts of a structure.

Woolly Sunflower

Gold is best: cf. the opening of Pindar, *Olympian* 1.

Harvest Lily

Coronaria is the adjective from *corona*, a crown or garland. The poem follows Clark's suggestion that the reference is to this plant's chaplet of flowers. However, Andy MacKinnon points out to me that, as a botanical term, *corona* refers to an appendix on a single flower (an extra whorl of parts between petals and stamens).

Staminodes are non-fertile stamens, lacking anthers. Nutrient-rich bulb: Haskin 9. Thousands in western Washington State: Lyons 299.

Cow Parsnip

Compound umbel: see "Spring Gold". Eaten by elk and bear: Craighead 126-27. Peeling the stems: Haskin 233; there is a photo by Robert D. Turner of the peeled stems in Turner, *Earth's Blanket* 118. On the phototoxin contained in the stems see Turner, *Nitinaht* 91-92. Special places for tastiest shoots: Nancy J. Turner and Barbara S. Efrat, *Ethnobotany of the Hesquiat Indians of Vancouver Island* (Victoria 1982) 61. Symonds: the references are to John Addington Symonds' poem "A Vista" (often sung as a hymn), which foresees "a loftier race" with "light of knowledge in their eyes"—a race which dares "All that may plant man's lordship firm/ On earth and fire, and sea, and air."

Part 3: Forest

Skunk Cabbage

Bears: Penn 20; noted already by Aristotle with reference to ἄρον (probably *Arum italicum*) in *History of Animals* 600b11. Comparison to ginger, Turner *Food Plants* 38; to pepper, Turner, *Nitinaht* 78. Heat breaks or rearranges crystals, Craighead 12. Poultices; Turner, *Nitinaht* 78; used when skin gone, Turner in a lecture.

Salmonberry

Chilcotin legend: Penn 102; Turner, *Earth's Blanket* 58, recounts a similar story concerning Raven and the dispersal of soapberry. Shoots eaten: Turner, *Food Plants* 126-27. 'Showy' is the meaning of *spectabilis.* Berries look like eggs: Penn 100. Sign of sockeye migration for Nuu-Chah-Nulth: Trevor C. Lantz and Nancy J. Turner, "Traditional Phenological Knowledge of Aboriginal Peoples in British Columbia," *Journal of Ethnobiology* 23 (2003) 263-86 at 265.

Thrush: Turner *Nitinaht* 124; Lantz and Turner (above) 266, 268; R. Wayne Campbell et al., *The Birds of British Columbia* (Vancouver 1990-2001) 3.394, 399. Story of Raven and salmonberry bird: recorded by Turner, *Earth's Blanket* 86-88, from Elsie Claxton and Violet Williams.

Red-flowering Currant

More than paid for Douglas' expedition: Clark 1976 229.

Orange Honeysuckle

Lack of scent: Clark 1976 492. Damage to host tree: ibid.

Western Trillium

Seed dispersal by ants (not unique to Trillium): Pojar & MacKinnon 102. Curse on pickers: Hitchcock et al. 1955-69, 1.807.

Calypso Orchid

For Calypso the nymph, see Homer's *Odyssey* Book 5; for Circe, Book 10.

Salal

This poem addresses changing human relationships to wilderness, through the specific example of salal. Wreck of the Valencia: http://www.historylink.org/essays/output.cfm?file_id=7382.
Salal cakes: Turner, Food Plants 77-78. Leaves for hunger: Turner, Nitinaht 105-06. John Scouler: Clark 1976 376. Flowers hanging in parallel rows: well illustrated Kruckeberg 81.

Dogwood

Symptoms of fungal disease (anthracnose) affecting dogwoods: plantclinic.cornell.edu/FactSheets/dogwoodanthracnose. Mighty elms: *ramos annosaque bracchia tendit/ ulmus opaca ingens* (Vergil, *Aeneid* 6.282-83). Strife and Love: Empedocles' names (in Greek, *Eris* and *Eros*) for these two forces.

Twinflower

Many an Ausländer: Clark 1976 489.

Vanilla Leaf

Angel wings: this image from Nancy Turner, pers. comm.; an alternative image is butterfly wings, Haskin 119. Veins persist: Pojar & MacKinnon 312. Fragrance based on coumarin: Haskin 119-21; Moore 252-53.

Tiger Lily

Tepals: this term is used when there is little or no differentiation between petals and sepals in a particular species.

Devil's Club

'Weapon': this is the meaning of Greek *[h]oplon*, the source of the first half of *Oplopanax*. Uses of Devil's Club: Pojar & MacKinnon 82; Turner, *Plant Technology* 140.

Fireweed

Silk mixed with wool for spinning: Turner, *Plant Technology* 175-6.

Part 4: Mountains

Harebell

Catharine Parr Traill, *Canadian Wild Flowers* (Montreal 1869, republished Toronto 1972) 56. Variety of forms and colours: R.J. Farrer, *The English Rock-Garden* (London 1928) 1.194. Resilience: Clark 1976 505. Linnaeus: Clark 1976 505. Polanyi: the reference is to Michael Polanyi, *Personal Knowledge: Towards a Post-Critical Philosophy*, Chicago 1958.

Lupine

Jewel-like droplets of dew: Haskin 186.

Fan-leaf Potentilla

Checking sepals to distinguish Buttercup from Potentilla: Harvey Manning 65. Epidermal cells of petals presenting glossy or matt appearance: Clark 1976 161.

Marsh-marigold

Some of the sound-play here is due to Lyons 177 ("pokes its precocious way") and Harvey Manning 41 ("squishy-soppy" meadows, and marsh marigold "exuberantly ushering in springtime").

Spreading Phlox

Bright flame of the sun: *hêliou phoibêi phlogi*, [Aeschylus] *Prometheus Bound* line 22. Halfway to the ocean: Harvey Manning 79.

Penstemon

Penstemon so named because five stamens are unusual among Figworts: Brown 191-92. Barely ten weeks of summer intervene from snow to snow: Haskin 317.

Paintbrush

Legend of origin of Paintbrush: Brown 162-63, based on Mabel Burkholder, *Before the White Man Came*. Co-evolution of columbine and hummingbirds: Art Kruckeberg et al, *Best Wildflower Hikes: Washington* (Seattle 2004) 115. Moral disapproval of parasitism: Brown 162-63. Discussion of Castilleja's partial parasitism, and quotation from John Muir (here adapted for metre), in Kruckeberg op. cit. 73.

Pink Monkeyflower

The bee-guide dots, and the flattened ends of the pistil, are well seen in the photo at Davies 78. Sensitive lobes of stigma: Haskin 329. Natural fingerbowls: Craighead 173.

Queen-cup

The name 'Queen-cup' was given by Julia W. Henshaw, *Mountain Wild Flowers of Canada* (Toronto 1906) p. 125. It remained in that form until Hitchcock and Cronquist 1973 p. 689 altered it to 'Queen's Cup', a form that has persisted since. Objections to the name *Clintonia* were raised by Thoreau (*Journal*, June 12 1852), who appears not to have known that DeWitt Clinton was a botanist as well as Governor of New York State. Haskin 35 objects both to *Clintonia* and to 'Queen-cup.' Freshness of plant matches mountain air: Clark 1976 16.

Yellow Glacier Lily

They "impatiently follow up the retreating snow banks," George & Winifred Hardy, *Wild Flowers in the Rockies* (Saskatoon 1949) p. 14. Rough cultivation by grizzly bears: Lyons 226. First nations' use of fire in managing land for food species: Turner, *Earth's Blanket* 151-61.

A Riddle

The answer is White-flowered Rhododendron, *Rhododendron albiflorum*.

Broad-leaved Willowherb

"As mountain sunlight streams across many an alpine gravel-bar, the flowers of the Broad-leaved Fireweed glow with an almost incredible brilliance", Clark 1976 329.

Sitka Valerian

Harvey Manning: 43. *Phu*: Dioscorides 1.11 (probably pronounced *Foo* in his day). Odour and medicinal use of Valerian root: Moore 249-50.

Pink Mountain Heather

"Nobody knows" is a rhetorical exaggeration, though the connection is certainly tenuous. Linnaeus named a member of the heath family after the Greek princess Andromeda; in a famous doodle (reproduced by Bernhardt 154) he whimsically indicated a similarity between the princess, chained to a sea-rock, and the heath, growing on a mound in a bog. A later botanist, David Don, developed this tradition and gave the name of Cassiope, Andromeda's mother, to a larger 'mother' genus of heaths (Bernhardt 76, 155). Bees by the thousands, bells by the millions: Harvey Manning 81.

Coda: The Mountains

Ascent is easy: a riff on Vergil's *facilis descensus Averni* (*Aeneid* 6.126). Moves the sun and the other stars: Dante, *Paradiso* canto 33.145. Skyladder: Christie Harris, *Raven's Cry* (Seattle 1992) p.76. Live and have our being: *Acts* 17.28.

Bibliography

Bernhardt, Peter, *Gods and Goddesses in the Garden: Greco-Roman Mythology and the Scientific Names of Plants*, New Brunswick 2008.

Britten, J., *Dictionary of English Plant Names,* London 1886.

Brown, Annora, *Old Man's Garden*, Sidney B.C. 1954.

Clark, Lewis J., *Wild Flowers of British Columbia*, Sidney B.C. 1973.

———(ed. John G. Trelawny), *Wild Flowers of the Pacific Coast*, Sidney B.C. 1976.

Craighead, John J.; Frank C. Craighead, Jr.; Ray J. Davis, *A Field Guide to Rocky Mountain Wildflowers*, Boston 1963.

Davies, Adrian, *Wild Flowers of Canada*, Godalming, U.K. 1989.

Elliott, Dave, Sr. (ed. Janet Poth), *Salt Water People: A Resource Book for the Saanich Native Studies Program*, Saanich 1983.

Garman, Eric H., *Pocket Guide to the Trees and Shrubs of British Columbia*, B.C. Forest Service Publication B 28, 3rd edn 1963.

Hardy, George A. & Winifred V., *Wild Flowers in the Rockies*, Saskatoon 1949.

Haskin, Leslie L., *Wild Flowers of the Pacific Coast*, Portland OR 1967.

Hitchcock, C.L.; A. Cronquist; M. Ownbey; J.W. Thompson, *Vascular Plants of the Pacific Northwest.* 5 vols. Seattle & London 1955-69.

Kruckeberg, Arthur R., *Gardening with Native Plants of the Pacific Northwest*, Vancouver & Toronto 1982 (2nd revised edition 2000).

Lyons, C.P., *Trees, Shrubs and Flowers to Know in British Columbia*, Toronto 1952.

———, *Shrubs and Flowers to Know in Washington,* Toronto 1956.

———and Bill Merilees, *Shrubs and Flowers to Know in British Columbia and Washington,* Vancouver 1995.

Manning, Harvey, Mountain Flowers of the Cascades and Olympics, Seattle 2nd edn. 2002.

Moore, Michael, *Medicinal Plants of the Pacific West*, Santa Fe 1993.

Penn, Briony, *A Year on the Wild Side*, Victoria 1999.

Pojar, Jim, and Andy MacKinnon, *Plants of Coastal British Columbia*, Vancouver 1994.

Quattrocchi, Umberto, *CRC World Dictionary of Plant Names*, 4 vols, Boca Raton 1999

Turner, Nancy J., *Food Plants of Coastal First Peoples* (Royal BC Museum Handbook), Victoria 1995.

———, *Plant Technology of First Peoples in British Columbia* (Royal BC Museum Handbook), Victoria 1998.

———, *The Earth's Blanket: Traditional Teachings for Sustainable Living*, Vancouver 2005.

———and Marcus A.M. Bell, "The Ethnobotany of the Coast Salish Indians of Vancouver Island," *Economic Botany* 25 (1971) 63-104.

———and Adam F. Szczawinski, *Edible Wild Fruits and Nuts of Canada*, Ottawa 1979.

———et al., *Ethnobotany of the Nitinaht Indians of Vancouver Island*, Victoria (Occasional Papers of the British Columbia Provincial Museum No. 24) 1983.

Ward-Harris, J., *More Than Meets the Eye: The Life and Lore of Western Wildflowers*, Toronto 1983.

Index

Author Information

John Fitch was formerly Professor and Chair of the Department of Greek and Roman Studies at the University of Victoria, and is now Professor Emeritus. His publications since retirement include a two-volume edition and translation of Seneca's dramas, and a translation of Palladius' *The Work of Farming*. For many years he lived on a farm on the Saanich Peninsula, where he grew fruit and raised sheep with his wife and daughter.

Bonnie Moro has a BA in psychology and a BFA in sculpture, in addition to undergraduate studies in botany and graduate studies in philosophy. Since childhood she has enjoyed drawing and painting plants as a means of understanding their detailed structure, and responding to it. She has prepared illustrations for the Cactus and Succulent Information Exchange and for the Rock and Alpine Club and the Rhododendron Society of Victoria.

Made in the USA
Monee, IL
04 November 2026

67377626R10106